The Masting of American Merchant Sail in the 1850s

Also by William L. Crothers

American-Built Packets and Freighters of the 1850s: An Illustrated Study of Their Characteristics and Construction
(McFarland, 2013)

The Masting of American Merchant Sail in the 1850s

An Illustrated Study

WILLIAM L. CROTHERS

McFarland & Company, Inc., Publishers
Jefferson, North Carolina

LIBRARY OF CONGRESS CATALOGUING-IN-PUBLICATION DATA

Crothers, William L., 1912–
The masting of American merchant sail in the 1850s :
an illustrated study / William L. Crothers.
p. cm.
Includes bibliographical references and index.

ISBN 978-0-7864-9399-9 (softcover : acid free paper) ♾
ISBN 978-1-4766-1569-1 (ebook)

1. Masts and rigging—United States—19th century.
2. Square-riggers—United States—Design and construction.
3. Sailing ships—United States—Design and construction.
4. Sailing ships—United States—History—19th century.
5. Merchant ships—United States—History—19th century.
I. Title.

VM531.C76 2014 623.822'4097309034—dc23 2014020475

BRITISH LIBRARY CATALOGUING DATA ARE AVAILABLE

On the cover: Clipper ship *Great Republic*, James E. Buttersworth, 1853
(© 2014 PicturesNow)

Printed in the United States of America

McFarland & Company, Inc., Publishers
Box 611, Jefferson, North Carolina 28640
www.mcfarlandpub.com

In remembrance of
my wife, Marjorie—
constant research companion
and ardent supporter.
We grew in the same garden
but she withered too soon
and I blossomed too late.

Acknowledgments

The subject of this book, the masting of American merchant sail in the 1850s, is obviously of rather restricted scope. In general it appears to be the by-product of other phases of sailing ship production. Any book on the subject requires the input of many individuals and institutions. Over a long period of time such information is gradually accumulated. If fortune is favorable a book emerges that successfully brings all the diverse pieces together to form a comprehensive picture.

When this has been accomplished it is time to sit back and remember those whose contributions have made it all possible and render recognition in, at least, some small way.

My sincere thanks and gratitude are herewith extended to those who, over the years, have contributed in some measure to the knowledge and sources of information that allowed it to become a feasible project.

E. Ann Wilcox, former librarian of the Philadelphia Maritime Museum (now the Independence Seaport Museum), who in that capacity furthered the progress of the work in many ways.

My late wife, Marjorie, my research assistant on the many field trips undertaken to ferret out and collect the many hidden bits of information that were brought to light.

Also, Ann Walsh, then library assistant, Webb Institute of Naval Architecture, Glen Cove, New York, and Paul J. O'Pecko, then reference librarian, Mystic Seaport Museum, Mystic, Connecticut.

Michael Costagliola, Sea Cliff, New York, and Robert and Grisel Leavitt, Miami, Florida.

Finally, the facilities of the Library of Congress and the National Archives, Washington, D.C.

To each of the above I express my sincere gratitude.

Table of Contents

Hulls are not ships until their
masts are stepped.—Sophocles

Preface

In the early 1990s, after many years of research and study on the subject of American merchant sailing vessels, it occurred to me that there might be room on bookshelves and in libraries for a book covering specifically the structure that was common to all these ships. At the time, apparently, in spite of all that had been written, no writer had committed to this subject.

Accordingly I decided to rectify this omission, selecting a time frame of the 1850s which were the glory years of American clippers. These vessels, along with the usual packets and freighters that were routinely built, presented a formidable pool of information. It was immediately apparent that this was a theme too large to fit conveniently between the covers of a single volume. This was especially true if one were to include the masts of ships as part of their structure.

Dividing the contents into separate parts was the obvious solution. This was done in the following manner: one volume was devoted to clippers; another volume covered packets and freighters; and this, the third volume, covers masting for all the ships.

The volumes could be considered a trilogy, each standing alone and complete in itself, but the three together form a related and connecting theme.

The first volume, *The American-Built Clipper Ship, 1850–1856*, was completed and published in 1996. Research material existed in great quantity due to the unprecedented popularity of these vessels.

The second volume, *American-Built Packets and Freighters of the 1850s*, was not favored with the quantity and availability of original data which accompanied the clippers. Consequently this volume was not published until 2013.

The final volume (this one) has been handicapped by the absence of original material dealing with the intimate details of mast making. American sources, such as builders' plans and newspaper articles, many times listed tables which contained the dimensions of masts but there was never a description of the internal construction of built-up masts.

However, there are two books, both British, that contain the information which was otherwise so difficult to find. The principal one of these is *A Treatise on Masting Ships and Mast Making* by John Fincham, originally published in 1829 with a second edition in 1843 and a third in 1854. This last edition contains all the details and instructions needed to build single stick and made masts and is the basis for the contents of this book.

The second reference that is available is a small, pocket-size volume titled *Masting, Mast-Making and Rigging of Ships, 1856* by Robert Kipping. Much of the contents of this book are abbreviated forms of Fincham's information.

Fincham's book was written specifically to update masting data that were included in David Steel's book *Elements of Mastmaking, Sailmaking and Rigging* which originally appeared in 1794.

This, the third volume of my trilogy, using text and a series of illustrations that are fashioned in a manner appropriate for the book and the period it embraces, completes my story of American wooden merchant sail in the 1850s.

Introduction

In the mid–1920s, in the United States, an interest in the commercial and naval sailing vessels that had been built during our relatively brief history began to take form. Books on the subject appeared in abundance.

Experts such as Charles G. Davis, naval architect, who was employed by the U.S. government to build wooden vessels in World War I; Howard I. Chapelle, naval architect; Carl C. Cutler, historian; Gordon Grant, artist; and others authored books on the subject.

Capt. E. Armitage McCann formed an organization of loosely connected clubs which were established in cities across the country. This body was named the Ship Model Makers Club and in February 1929 it published the first issue of its journal, *The Ship Modeler*. This organization and its journal flourished until 1934. The final issue of the journal was published in October-November 1933 and the club dissolved in 1934.

In the meantime other clubs had formed and many authors had brought forth additional books on the subject.

Sailing ships were written about in every conceivable way. Their lines were scrutinized, hull forms studied, scantlings examined, masting and rigging analyzed. There was not an item that did not undergo intensive study.

Many of the old-time shipyards left behind voluminous records of the vessels they built. There was information enough to replicate some of the ships in their entirety, including the hull timbers, fastenings, fittings, masting, rigging and sails.

Available for study and research were tables of offsets, lines plans, structural plans, sail plans and masting tables. The masting tables consisted of a list of all the masts and spars that would appear in a vessel, complete with the lengths, diameters, headings and other data pertinent to a complete mast.

Up to this point it appeared that every facet of the wooden sailing ship had been exhaustively covered. It was not quite so.

In the course of building increasingly larger vessels it became more and more necessary to increase the size of masts and sails to power them. This was done by felling larger and larger trees to fill the need. Ultimately the diameters of the largest masts exceeded the diameter of the largest trees. (The most extreme case occurred with *Great Republic*, in which vessel the mainmast was 44 inches maximum diameter and 131 feet in length, all in a continuous unit.) The way to resolve the situation was to bundle smaller trees together into one stick of the desired size. Thus, the single tree mast was joined by the built or made mast.

This was where the completeness of the sailing ship picture failed. Since it was the external form of the complete mast that was important, the internal mass was relatively immaterial. However, the built-up mast had to be made. How was it done?

A system was devised whereby a squared central stick, called a core or spindle, was surrounded by various other squared sticks, each having its own specific identity, until the desired diameter of the mast was achieved. This assembly was then tightly bound together to form a solid unit.

In 1794, David Steel (British) wrote a treatise on mast making which covered the general subject and was, in essence, the authority at that time.

In 1829, John Fincham (British) published a work on masting because of the general need for an updated version of the subject.

In 1854, he published a new edition (the third edition) of the book, further enlarging on the contents to make them more relevant to the times.

In 1856, Robert Kipping (British) published a short, rather abbreviated book on the subject of mast making, often referring to Fincham's book for certain information.

The data in these three books comprise the knowledge from which this current volume has evolved.

There is no guarantee that the practices that were recorded in those books were adhered to in American shipbuilding. However, those practices had been put to use for a considerable length of time and had proven to be satisfactory. There is no compelling reason that American naval architects, shipbuilders and mastmakers would digress from a proven system.

In the appendix of this volume there are specific instances where it is stated that certain ships' masts were built or made. In a few instances the actual number of sticks in a

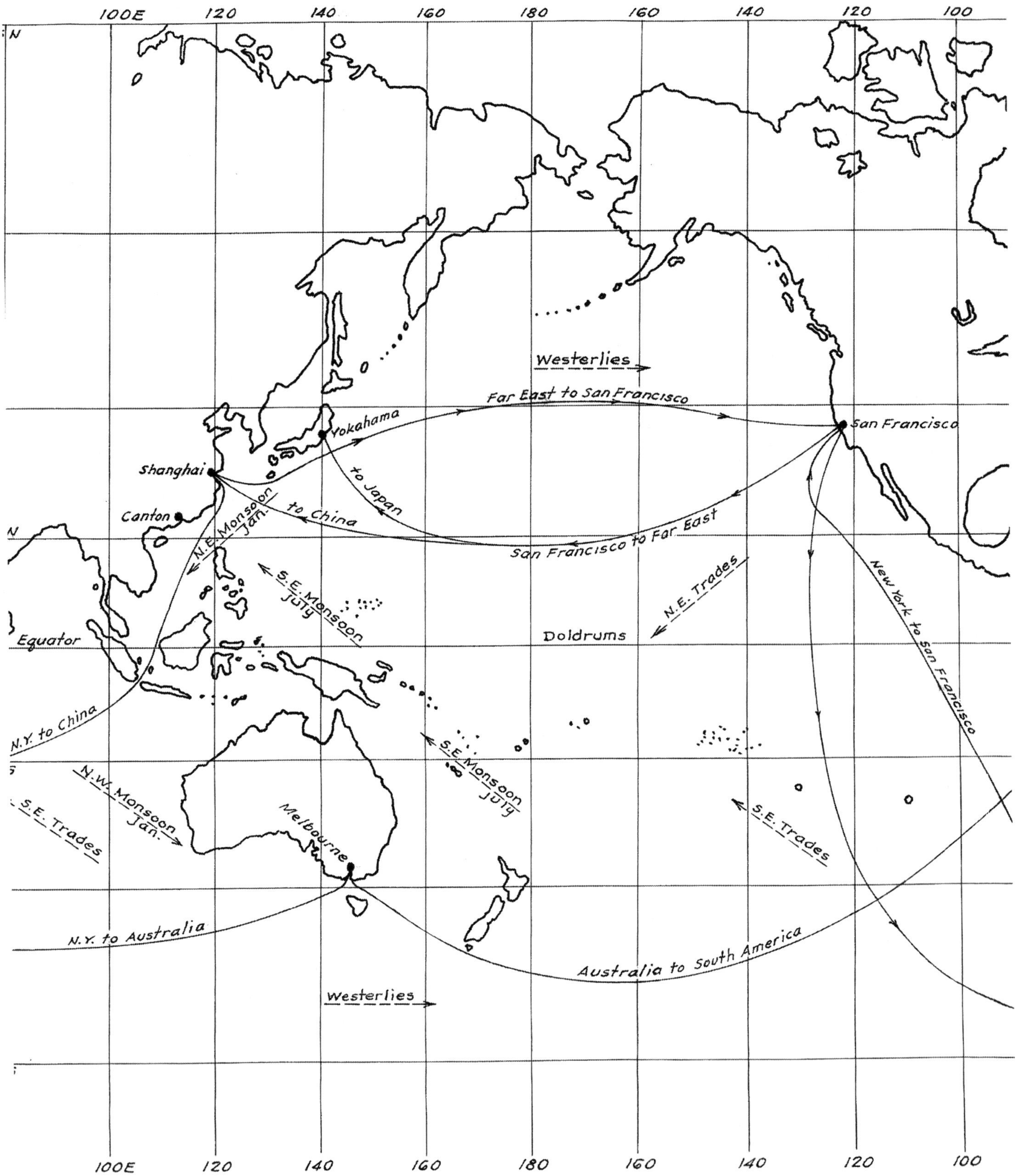

Above and opposite: **Routes used by American merchant sail in the 1850s.**

mast is recorded. Suffice it to say that, if nothing else, American mast makers could have used this information that had proven successful countless times over.

In any case, the ingenuity and craftsmanship that were required to produce a large made mast are both surprising and interesting and will open an unusual view of one of the lesser known details of the wooden sailing ship.

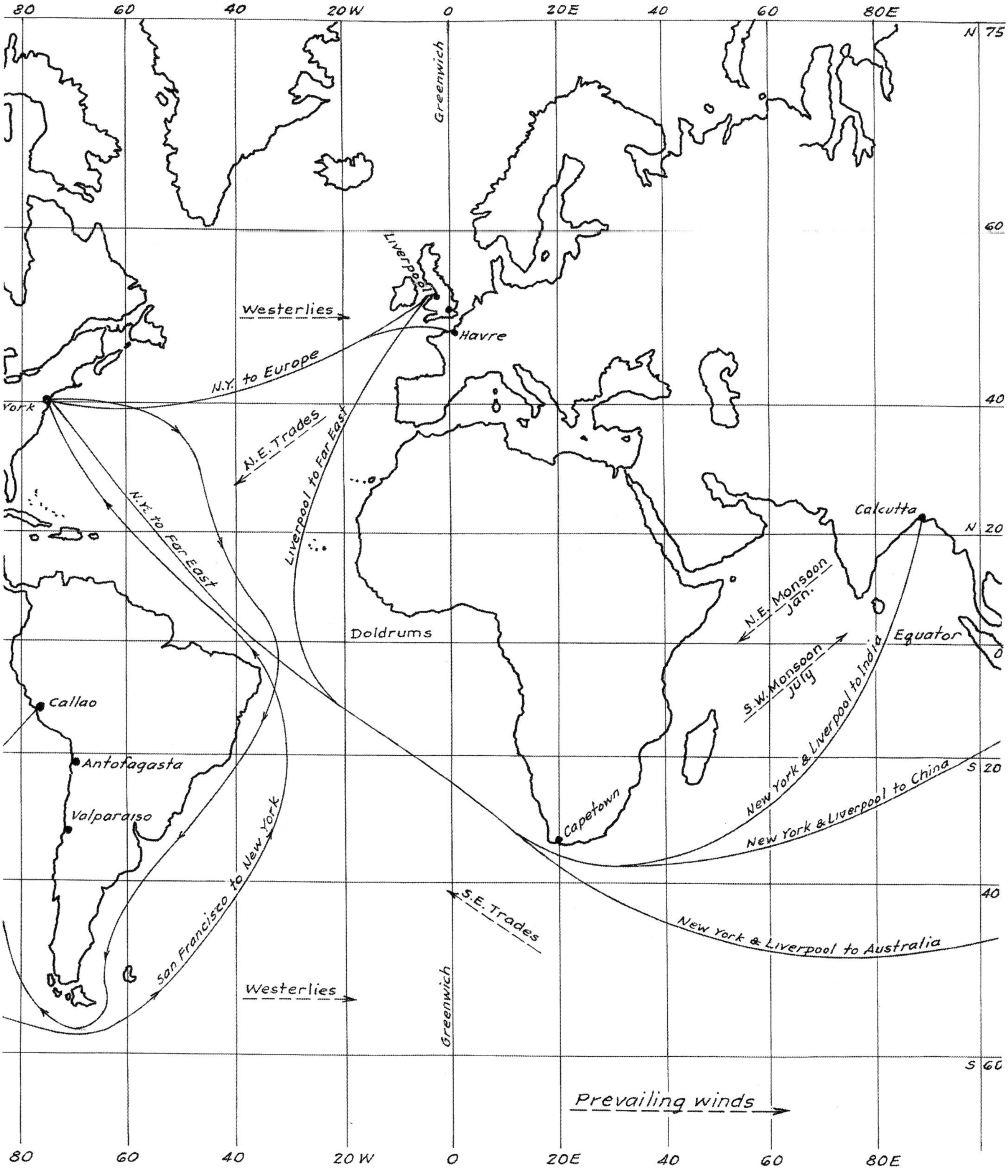

80
60
40
20 W
0
20E
40
60
80E
N 75
60
40
N 20
0
S 20
40
S 60
Greenwich
Liverpool
Westerlies
Havre
N.Y. to Europe
York
N.E. Trades
Liverpool to Far East
N.Y. to Far East
Calcutta
N.E. Monsoon
Jan.
Doldrums
Equator
S.W. Monsoon
July
New York & Liverpool to India
Callao
Antofagasta
Valparaiso
Capetown
New York & Liverpool to China
San Francisco to New York
S.E. Trades
New York & Liverpool to Australia
Westerlies
Greenwich
Prevailing winds

Alphabetical List of Vessels

Vessels are ship rigged unless otherwise noted.

Vessel	*Launch*	*Category*	*Signification of Name*
Abbott Lawrence	1855	packet	American manufacturer and statesman, born in Boston, 1792.
Alarm	1856	clipper	
Alice Tainter, bark	1856	freighter	
Alma	1855	freighter	A feminine name meaning "bountiful."
Amos Lawrence	1855	packet	American colonizer and philanthropist of Boston, 1814.
Amphitrite	1853	clipper	Greek goddess of the sea.
Andrew Jackson	1855	clipper	Seventh president of the United States. Launched as *Belle Hoxie*.
Antarctic	1850	packet	The region at or around the South Pole.
Antelope, barkentine	1855	freighter	A swift, hollow-horned, cud-chewing, deer-like animal.
Asa Eldridge	1856	clipper	Named in memory of the captain of the lost Collins steamer *Pacific*.
Aurelia, bark	1855	freighter	A feminine name derived from aurum (gold).
Aurora	1854	packet	Roman goddess of dawn.
Bald Eagle	1852	clipper	National bird in the United States coat of arms.
Barreda Brothers	1854	packet	In compliment to an eminent firm in Baltimore.
Belle, brig	1850	packet	A pretty woman or girl.
Belle, brig	1853	clipper	A pretty woman or girl.
Belle of the West	1853	clipper	
Benin, bark	1854	freighter	A (former) native kingdom of West Africa. The vessel was built for the Liverpool and West Africa trade.
Black Hawk	1856	clipper	
Blue Jacket	1854	clipper	A description of a seaman.
Bounding Billow, bark	1854	clipper	
Brenda, topsail schooner	1851	freighter	A feminine name derived from "brand," a sword. (Vessel is a duplicate of *Minna*.)
Cathedral	1855	packet	A large, imposing church.
Challenge	1851	clipper	
Challenger	1853	clipper	
Champion of the Seas	1854	clipper	
Charmer	1854	clipper	
Chatsworth	1854	packet	Named as a compliment to the seat of the English duke of Devonshire.
Clara Wheeler	1850	packet	A woman prominent in the 19th century westward expansion of the United States.
Climax	1853	clipper	
Coeur de Lion	1854	clipper	"Lion hearted," Richard I, king of England, 1189–1199.
Comet	1851	clipper	A heavenly body in the solar system, with a long, luminous tail.
Commodore Perry	1854	packet	Named after Matthew Calbraith Perry, U.S. naval officer who negotiated a trade treaty with Japan in 1854.
Constantine	1850	packet	Early Roman emperor.
Cornelius Grinnell	1850	packet	Member of a well-known shipbuilding family of the 1850s.
Dauntless	1852	clipper	
Dione	1857	freighter	Greek goddess of the Earth. Also, sometimes, goddess of the oak tree.
Donald McKay	1855	clipper packet	Ship designer and builder in Boston, 1845–1869.
Eagle Wing	1853	clipper	

Vessel	*Launch*	*Category*	*Signification of Name*
Edisto, bark	1851	packet	Possibly named after Edisto Island, South Carolina.
Edwin Forrest	1853	clipper	Famous American actor, 1804–1872.
Electric Spark	1855	clipper	
Ellen Foster	1852	clipper	The wife of her builder, Joshua T. Foster, Medford, Massachusetts.
Elvira	1855	freighter	A feminine name.
Emerald	1855	packet	A bright green precious stone.
Empress of the Sea	1853	clipper	Contemporary spelling of the name using "*Sea*" in lieu of "*Seas.*"
Endeavor	1856	clipper	
Eringo, bark	1853	clipper	Possibly named after the herb eryngium or eryngo. The candied roots were believed to possess aphrodisiac properties. The name was also freely spelled eringo.
Fair Wind	1855	clipper	
Fearless	1853	clipper	
Flying Arrow	1852	clipper	Built as *Flying Yankee.*
Flying Childers	1852	clipper	Famous undefeated English racehorse of the 1700s.
Flying Cloud	1851	clipper	
Flying Dutchman	1852	clipper	A Dutch sailor condemned to sail the seas until Judgement Day.
Flying Eagle	1852	clipper	
Flying Fish	1851	clipper	
Flying Mist	1856	clipper	
Galatea	1854	clipper	Greek mythological statue of a maiden, given life by Aphrodite after its sculptor, Pygmalion, fell in love with it.
Game Cock	1850	clipper	
Gazelle	1851	clipper	
Golden Fleece	1855	clipper	Fleece of gold guarded by a dragon and taken away by Jason and the Argonauts.
Golden Light	1853	clipper	
Good Hope	1855	freighter	Possibly named after the southern tip of Africa.
Great Republic, 4-mast bark	1853	clipper	A reference to the United States of America.
Great Republic, (rebuilt) 4-mast bark	1855	clipper	As rebuilt with reduced rig.
Halifax, bark	1855	clipper	Named after the capital of Nova Scotia.
Harry Bluff	1855	packet	
Henry Hill, bark	1856	clipper	
Herald of the Morning	1853	clipper	A reference to Eos (Greek) or Aurora (Roman), goddess of the morning.
Hesperus	1856	freighter	The evening star (Greek).
Hoogly	1851	clipper	Named after the navigable Houghly River, an arm of the Ganges, in India.
Hurricane	1851	clipper	
Invincible	1851	clipper	
J. Montgomery	1852	packet	
James Baines	1854	clipper	Senior member of the owners, James Baines & Co., Liverpool.
Japan	1854	packet	
John Bertram	1850	clipper	Well-known sea captain and merchant of Salem, Massachusetts.
John Gilpin	1852	clipper	A London draper whose horseback riding is celebrated in the poem "The Diverting History of John Gilpin" by William Cowper, English poet, 1731–1800.
John Land	1853	clipper	Named after Capt. John Land, once master of *Rainbow* and *Challenge.* (This vessel is an exact replica of *Winged Arrow.*)
John Stuart	1851	clipper	
Joseph Peabody	1856	clipper	Name of prominent shipowner of Salem, Massachusetts, in the early 1800s.
Joseph Walker	1850	packet	
Josephine, brigantine	1855	freighter	
Kremlin	1850	freighter	
Lady Suffolk	1855	packet	A titled lady of England.
Lamplighter, bark	1854	clipper packet	
Lightning	1854	clipper	
Mameluke	1855	clipper	Member of a military caste which ruled Egypt from 1250 to 1517.
Manhasset, schooner	1852	freighter	Name of a Long Island, New York, community. (Vessel is a duplicate of *Plandome.*)

Vessel	*Launch*	*Category*	*Signification of Name*
Mermaid, bark	1851	clipper	
Minna, topsail schooner	1851	freighter	Vessel is a duplicate of *Brenda*.
Moneynick, bark	1854	packet	Name of a place in Ireland where the forefathers of James Adger, of Charleston, South Carolina, one of her owners, lived.
Mystery	1854	clipper	
Nightingale	1851	clipper	Named after Swedish soprano Jenny Lind, 1820–1887
Noonday	1855	clipper	
Ocean Monarch	1856	freighter	
Orient	1852	packet	
Phantom	1852	clipper	
Plandome, schooner	1852	freighter	Name of a Long Island, New York, community. (Vessel is a duplicate of *Manhasset*.)
Plutarch	1856	packet	A Greek biographer and historian, 46–120.
Polar Star	1852	freighter	Named after Polaris, the North Star.
Queen of Clippers	1853	clipper	
Queen of the Seas	1852	clipper	
Quickstep, bark	1855	clipper	
Racehorse, bark	1850	clipper	Classed as a clipper by Capt. Arthur H. Clark. Classification refuted later by scholars.
Racer	1851	clipper	
Radiant	1853	clipper	
Red Jacket	1853	clipper	Popular name of Sagoyewatha, Seneca Indian chief, living at the time of the American Revolution.
Roebuck	1851	clipper	
Romance of the Sea	1853	clipper	A reference to Sir Walter Scott and James Fenimore Cooper.
Samuel Lawrence	1851	freighter	
Saracen	1854	clipper	A member of the early nomadic tribes of the Middle East.
Sea Bird, bark	1851	clipper	
Sea Witch	1846	clipper	
Seaman's Bride	1851	clipper	Named to complement the clipper *Seaman*, launched in 1850 by the same builders.
Shooting Star	1851	clipper	
Sierra Nevada	1854	clipper	Name changed from *King of the Forest* before launch.
Silver Star	1856	clipper	
Southern Cross	1851	clipper	A heavenly constellation.
Sovereign of the Seas	1852	clipper	Named after the original ship built at Woolwich, England, in 1637. While on the stocks was named *Enoch Train*. See ref. 1, edition of 25 May 1852. Name changed prior to launch.
Spitfire	1853	clipper	
Staffordshire	1851	clipper packet	English pottery district, source of much of Enoch Train's "Train Line" cargoes.
Stag Hound	1850	clipper	Contemporary spelling. (Two words.)
Star of Empire	1853	packet	A poetical designation. "Westward the Star of Empire takes its way," quoted from "Oration at Plymouth 1802" by John Quincy Adams.
Storm King	1853	clipper	
Sultana, bark	1850	clipper	Title of the wife, mother, sister or daughter of a sultan.
Surprise	1850	clipper	
Susan Howland	1856	freighter	
Sweepstakes	1853	clipper	
Sword Fish	1851	clipper	
Syren	1851	clipper	
Thatcher Magoun	1856	clipper	Senior member of the owners, T. Magoun & Son, Boston, Massachusetts.
Trieste, bark	1857	freighter	Seaport on the Gulf of Trieste, Italy.
Union	1850	freighter	
Vitula	1855	freighter	
Volante, brig	1852	freighter	
War Hawk	1855	clipper	
Water Witch	1853	clipper	
Webster	1853	packet	A tribute to Daniel Webster.

Vessel	*Launch*	*Category*	*Signification of Name*
West Wind	1853	clipper	
Westward Ho	1852	clipper	
Wild Pigeon	1851	clipper	
Wildfire, bark	1853	clipper	
Winchester	1851	freighter	Named after Col. W.P. Winchester, of Boston, Massachusetts.
Winfield Scott	1851	packet	A great American general.
Winged Arrow	1852	clipper	Built as a duplicate of the clipper *John Land*.
Winged Racer	1852	clipper	
Witch of the Wave	1851	clipper	
Witch of the Wave	1856	freighter	
Witchcraft	1850	clipper	
Young America	1853	clipper	A reference to the United States of America.
Young Turk	1856	clipper	

Chronological List of Vessels

All vessels are ship rigged unless otherwise noted. All dimensions are in feet and inches. oa = Length overall. od = Length on deck. pp = Length between perpendiculars. * Length of this dimension is not specifically stated.

Vessel & Ref. #	*Launch*	*Designer*	*Builder*	*Original Owner*
Sea Witch #29 192oa × 34 ×19 908 tons 2 decks	1846 December	W. Griffiths, New York	Smith & Dimon, New York	Howland & Aspinwall, New York
Clara Wheeler #1 184oa × 37-9 × 26-9 1000 tons 3 decks	1850 January	—	John Taylor, Medford, Mass.	Bramhall & Howe, Boston
Belle, brig #1 95od × 24 × 14 206 tons 1 deck	1850 June	—	Bigelow & Co., Cornwallis, Nova Scotia	Clark & Jones, Boston
Kremlin, bark #1 130od × 29 × 19 475 tons 2 decks	1850 June	Capt. A.A. Frazer, Medford, Mass.	Capt. A.A. Frazer, Medford, Mass.	A New York group
Racehorse, bark #1 128oa × 30-6 × 16 514 tons 1 deck and hold beams partially decked	1850 June	Samuel Hall, East Boston (sometimes attributed to Samuel Hartt Pook)	Samuel Hall, East Boston	J.M. Forbes, Boston
Sultana, bark #1 121od × 28-4 × 15 434 tons 1 deck and hold beams	1850 June	Donald McKay, East Boston	Donald McKay, East Boston	Edw. Lamb & Brother, Boston
Joseph Walker #21, 24 180d × 39-4 × 23 1500 tons 2 decks	1850 August	Wm. H. Webb, New York	Wm. H. Webb, New York	Sam'l Thompson & Nephew, New York
Antarctic #1 173od × 38-6 × 23-6 1115 tons 2 decks	1850 September	Donald McKay, East Boston	Donald McKay, East Boston	Zerega & Co., New York
Constantine #1 205oa × 38-6 × 23-9 1200 tons 2 decks and hold beams	1850 September	George Raynes, Portsmouth, N.H.	George Raynes, Portsmouth, N.H.	Grinnell & Minturn, New York

Vessel & Ref. #	*Launch*	*Designer*	*Builder*	*Original Owner*
Surprise #1 190oa × 39 × 22 1262 tons 2 decks	1850 October	Samuel H. Pook, Boston	Samuel Hall, East Boston	A.A. Low & Bro., New York
Union #1 151pp × 34-6 × 21 700 tons 2 decks	1850 November	Samuel Lapham, Medford, Mass.	Samuel Lapham, Medford, Mass.	Daniel C. Bacon, Boston
Game Cock #1 200oa × 40 × 22 1891 tons 2 decks	1850 December	Samuel H. Pook or Samuel Hall (disputed)	Samuel Hall, East Boston	Daniel C. Bacon, Boston
John Bertram #1 190oa × 37 × 20 1050 tons 2 decks	1850 December	—	Ewell & Jackson, East Boston	Glidden & Williams, et al., Boston
Stag Hound #1 226oa × 39-8 × 21 1534 tons 2 decks	1850 December	Donald McKay, East Boston	Donald McKay, East Boston	George B. Upton, Sampson & Tappan, Boston
Witchcraft #1 193oa × 39-4 × 22 1311 tons 2 decks	1850 December	Samuel H. Pook, Boston	Paul Curtis, Chelsea, Mass.	Rogers & Pickman, Salem, Mass.
Cornelius Grinnell #30 182* × 36-5 × 23-6 1118 tons 2 decks	1850 December	Donald McKay, East Boston	Donald McKay, East Boston	Grinnell & Minturn, New York
Gazelle #6 181od × 40 × 21 1244 tons 2 decks	1851 January	Wm. H. Webb, New York	Wm. H. Webb, New York	Taylor & Merrill, New York
Shooting Star #1 171oa × 35 × 18-6 903 tons 2 decks	1851 February	Capt. John Wade, Boston	James O. Curtis, Medford, Mass.	Reed, Wade & Co., Boston
Mermaid, bark #1 145oa × 29-3 × 15 533 tons 1 deck and hold beams partially decked	1851 March	Samuel Hall, East Boston	Samuel Hall, East Boston	Hall, Gassell & May, Boston
Southern Cross #1 175oa × 36 × 21 938 tons 2 decks	1851 March	E. & H.O. Briggs, South Boston	E. & H.O. Briggs, South Boston	Baker & Morrill, Boston
Edisto, bark #1 126od × 26-4 × 16 366 tons 2 decks	1851 April	—	Hayden & Cudworth, Medford, Mass.	Adger & Caldwell, Charleston, S.C.; Lombard & Co., Capt. N.L. Kendrick, et al., Boston
Flying Cloud #1 220oa × 41 × 21-6 1783 tons 2 decks	1851 April	Donald McKay, East Boston	Donald McKay, East Boston	Grinnell & Minturn, New York

Vessel & Ref. #	*Launch*	*Designer*	*Builder*	*Original Owner*
Witch of the Wave #1 220oa × 40 × 21 1498 tons 2 decks	1851 April	George Raynes, Portsmouth, N.H.	George Raynes, Portsmouth, N.H.	Glidden & Williams, et al., Boston
Challenge #1 252-6oa × 43 × 25-6 2006 tons 3 decks	1851 May	Wm. H. Webb, New York	Wm. H. Webb, New York	N.L. & G. Griswold, New York
Syren #3 189oa × 36 × 22 1064 tons 2 decks	1851 May	—	John Taylor, Medford, Mass.	Silsbee & Pickman, Salem, Mass.
Nightingale #3 185od × 36-6 × 20 1070 tons 2 decks	1851 June	Sam'l. Hanscom, Jr. & Capt. F.A. Miller, Portsmouth, N.H.	Sam'l Hanscom, Jr., Portsmouth, N.H.	Davis & Co., et al., Portsmouth, N.H.
Racer #3 207oa × 42-6 × 28 1650 tons 3 decks	1851 June	—	Currier & Townsend, Newburyport, Mass.	David Ogden, et al., New York
Seaman's Bride #2, 6 152od × 31-6 × 17-6 668 tons 2 decks	1851 June	—	R. & E. Bell, Baltimore	Thomas J. Handy & Co., New York
Staffordshire #1 240oa × 41 × 29 1817 tons 3 decks	1851 June	Donald McKay, East Boston	Donald McKay, East Boston	Enoch Train & Co., Boston
Comet #2,4 241oa × 41-4 × 22-2 1836 tons 2 decks	1851 July	Wm. H. Webb, New York	Wm. H. Webb, New York	Bucklin & Crane, New York
Wild Pigeon #1 184oa × 36-4 × 20 996 tons 2 decks	1851 July	George Raynes, Portsmouth, N.H.	George Raynes, Portsmouth, N.H.	Olyphant & Co., New York
Winfield Scott #1 194oa × 38-3 × 30 1240 tons 3 decks	1851 July	John Glidden, Esq., Newcastle, Maine	John Glidden, Newcastle, Maine	John Glidden, Newcastle, Maine
Invincible #6 235oa × 42-10 × 25-6 1769 tons 3 decks	1851 August	Wm. H. Webb, New York	Wm. H. Webb, New York	James W. Phillips, New York
Flying Fish #1 220oa × 39-6 × 22 1505 tons 2 decks	1851 September	Donald McKay, East Boston	Donald McKay, East Boston	Sampson & Tappan, Boston
Samuel Lawrence #1 185oa × 38 × 27-6 1040 tons 3 decks		Paul Curtis, Medford, Mass.	Paul Curtis, Medford, Mass.	Andrew T. Hall, Esq., Boston

Vessel & Ref. #	*Launch*	*Designer*	*Builder*	*Original Owner*
Sword Fish #1 170od × 36-6 × 20 1036 tons 2 decks	1851 October	Wm. H. Webb, New York	Wm. H. Webb, New York	Barclay & Livingston, New York
Brenda, topsail schooner #1 112od × 27-6 × 12-6 300 tons, 1 deck (Duplicate of *Minna*)	1851 October	George Raynes, Portsmouth, N.H.	George Raynes, Portsmouth, N.H.	R.B. & J.M. Forbes, Portsmouth, N.H.
Hurricane #6 230oa × 40 × 22 1608 tons 2 decks	1851 October	—	Isaac C. Smith, Hoboken, N.J.	C.W. & H. Thomas, New York
John Stuart #6 228-6oa × 43 × 28 1670 tons 3 decks	1851 October	Perrine, Paterson & Stack, Williamsburgh, New York	Perrine, Paterson & Stack, Williamsburgh, New York	B.A. Mumford, J. Smith, et al., New York
Minna, topsail schooner #1 112od × 27-6 × 12-6 300 tons 1 deck (Duplicate of *Brenda*)	1851 October	George Raynes, Portsmouth, N.H.	George Raynes, Portsmouth, N.H.	R.B. & J.M. Forbes, Portsmouth, N.H.
Winchester #1 215oa × 41-6 × 29 1480 tons 3 decks	1851 October	—	Ewell & Jackson, East Boston	Weld & Baker, Boston
Sea Bird, bark #1 115oa × 26 × 14 325 tons 1 deck and hold beams partially decked	1851 October	Samuel Hall, East Boston	Samuel Hall East Boston	Lombard & Co., et al., Boston
Hoogly #1 200oa × 39 × 25 1303 tons 2 decks and hold beams partially decked	1851 December	Samuel Hall, East Boston	Samuel Hall, East Boston	D.C. Bacon & Sons, Boston
Roebuck #1 170oa × 33 × 22 816 tons 2 decks	1851 December	—	Bourne & Kingsbury, Kennebunk, Maine	Nichols & Curtis, Boston
Dauntless #1 185oa × 33 × 21-6 800 tons 2 decks	1852 January	W.W. Goddard, Esq., Boston	Benj. F. Delano, Medford, Mass.	W.W. Goddard, Esq., Boston
Polar Star #1 160oa × 33 × 23 670 tons 2 decks	1852 January	—	J.T. Foster, Medford, Mass.	Pearson & Co., Boston
Manhasset, schooner #24, 26 101od × 25-7 × 10-9 250 tons 1 deck (Duplicate of *Plandome*)	1852 February	Wm. H. Webb, New York	Wm. H. Webb, New York	Bernard Flanner, New York

Vessel & Ref. #	*Launch*	*Designer*	*Builder*	*Original Owner*
Plandome #24, 26 101od × 25-7 × 10-9 250 tons 2 decks	1852 February	Wm. H. Webb, New York	Wm. H. Webb, New York	Sam'l L. Mitchill, New York
Ellen Foster #1 180oa × 37 × 24 996 tons 2 decks	1852 April	—	Joshua T. Foster, Medford, Mass.	J. & A. Tirrell, Boston
Sovereign of the Seas #1 265oa × 44 × 23-6 2421 tons 2 decks and hold beams	1852 June	Donald McKay, East Boston	Donald McKay, East Boston	Donald McKay, Boston
Winged Arrow #1 183oa × 36 × 22 1052 tons 2 decks	1852 July	E. & H.O. Briggs, South Boston	E. & H.O. Briggs, South Boston	Baker & Morrill, Boston
J. Montgomery #1 181oa × 36 × 23-3 893 tons 2 decks	1852 August	Samuel Badger, Portsmouth, N.H.	Samuel Badger, Portsmouth, N.H.	S.C. Thwing, Portsmouth, N.H.; R.D. Shepherd, Esq., Virginia; Capt. John Davis, Portsmouth, N.H.
Flying Dutchman #1 200oa × 38-6 × 21-6 1257 tons 2 decks	1852 September	Wm. H. Webb, New York	Wm. H. Webb, New York	George B. Daniels, et al., New York
John Gilpin #1 205oa × 37 × 22 1084 tons 2 decks	1852 September	Samuel Hall, East Boston	Samuel Hall, East Boston	Pierce & Hunnewell, Boston
Queen of the Seas #1 214oa × x39 × 22 1356 tons 2 decks	1852 September	—	Paul Curtis, Medford, Mass.	Glidden & Williams, Boston
Westward Ho #1 220oa × 40-6 × 23-6 1600 tons 2 decks	1852 September	Donald McKay, East Boston	Donald McKay, East Boston	Sampson & Tappan, Boston
Orient #1 215oa × 41-6 × 29-6 1560 tons 3 decks	1852 October	George Raynes, Portsmouth, N.H.	George Raynes, Portsmouth, N.H.	Spofford, Tileston & Co., et al., New York
Bald Eagle #1 225oa × 41-6 × 22-6 1705 tons 2 decks	1852 November	Donald McKay, East Boston	Donald McKay, East Boston	George B. Upton, Boston
Flying Childers #1 195oa × 36-4 × 22-6 1125 tons 2 decks	1852 November	Samuel Hall, East Boston	Samuel Hall, East Boston	J.M. Forbes & Cunningham Bros., Boston
Winged Racer #1 226oa × 42-6 × 23 1767 tons 2 decks	1852 November	Samuel H. Pook, Boston	Rob't E. Jackson, East Boston	Seccomb & Taylor, Boston

Vessel & Ref. #	*Launch*	*Designer*	*Builder*	*Original Owner*
Flying Arrow #1 170-8od × 37-10 × 23-4 1092 tons 2 decks	1852 December	—	George Dunham, Frankfort, Maine	Manning, Stanwood & Co., Boston
Flying Eagle #1 195oa × 37 × 23 1092 tons 2 decks	1852 December	—	Wm. Hitchcock, Damariscotta, Maine	F. Nickerson, J. Beal, et al., Boston
Phantom #1 200oa × 37-11 × 21-6 1300 tons 2 decks	1852 December	Samuel Lapham, Medford, Mass., & John F. Lodge, Boston	Samuel Lapham, Medford, Mass.	Crocker & Sturgis, et al., Boston
Volante #24, 27 brig 112od × 26-6 × 11-3 307 tons 1 deck	1852 December	Wm. H. Webb, New York	Wm. H. Webb, New York	Schiff Brothers & Co., New York
Empress of the Sea #1 240oa × 43 × 27 2200 tons (est.), 2 decks, hold beams partially decked	1853 January	Donald McKay, East Boston	Donald McKay, East Boston	William Wilson & Son, Baltimore
Golden Light #1 193oa × 36 × 22-6 1150 tons 2 decks	1853 January	E. & H.O. Briggs, South Boston	E. & H.O. Briggs, South Boston	James Huckins & Sons, Baltimore
Mystery #1 196oa × 36 × 22-6 1150 tons 2 decks	1853 January	Samuel Hall, East Boston	Samuel Hall, East Boston	Crocker & Sturgis, Boston
Radiant #1 210oa × 40 × 24-4 1300 tons 2 decks	1853 January	—	Paul Curtis, East Boston	Baker & Morrill, Boston
Storm King #1 216oa × 39 × 23 1400 tons 2 decks	1853 February	—	John Taylor, Chelsea, Mass.	Snow & Rich, Boston
Belle of the West #1 182oa × 35 × 20-6 936 tons 2 decks	1853 March	Samuel H. Pook, Boston	Shiverick Bros., East Dennis, Mass.	Glidden & Williams, Boston
Climax #1 180od × 36 × 22-9 1051 tons 2 decks	1853 March	—	Hayden & Cudworth, Medford, Mass.	Howes & Crowell, Boston
John Land #1 183oa × 36 × 22 1052 tons 2 decks	1853 March	E. & H.O. Briggs, South Boston	E. & H.O. Briggs, South Boston	Baker & Morrill, Boston
Queen of Clippers #1 258oa × 44-6 × 24 2300 tons 2 decks	1853 March	—	Rob't E. Jackson, East Boston (bought from Seccomb & Taylor before launch)	Zerega & Co., New York

Vessel & Ref. #	*Launch*	*Designer*	*Builder*	*Original Owner*
West Wind #1 180od × 36-6 × 24 1071 tons 2 decks	1853 March	—	Joshua T. Foster, Medford, Mass.	J. & A. Tirrell, Boston
Star of Empire #1 230od × 43 × 27-6 2050 tons 3 decks	1853 April	Donald McKay, East Boston	Donald McKay, East Boston	Enoch Train & Co., Boston
Wildfire, bark #1 138oa × 28 × 10-6 350 tons 1 deck and high quarter deck	1853 April	Simeon McKay, Boston	Simeon McKay, Boston	A.L. Payson, Boston
Young America #2 243od × 43-2 × 26-9 1961 tons 3 decks	1853 April	Wm. H. Webb, New York	Wm. H. Webb, New York	George B. Daniels, New York
Water Witch #1 192oa × 38-3 × 21 1204 tons 2 decks	1853 May	—	Fernald & Pettigrew, Portsmouth, N.H.	Stephen Tilton, et al., Boston
Amphitrite #1 221od × 41 × 26-3 1687 tons 2 decks and hold beams partially decked	1853 June	Samuel Hall, East Boston	Samuel Hall, East Boston	Samuel Hall, East Boston
Belle, brig #1 95od × 28 × 11-6 213 tons 1 deck	1853 June	—	C.W. Conners, Cornwallis, Nova Scotia	Clark & Jones, Halifax, Nova Scotia
Sweepstakes #29 216-4od × 41-6 × 22 1735 tons 2 decks	1853 June	Daniel Westervelt, New York	Daniel & A. Westervelt, New York	Grinnell & Minturn, New York
Fearless #29 191* × 36-5 × 22 1184 tons 2 decks	1853 July	Samuel H. Pook, Boston	A. & G.T. Sampson, East Boston	William F. Weld & Co., Boston
Spitfire #1 224oa × 40 × 23 1550 tons 2 decks	1853 September	—	James Arey & Co., Frankfort, Maine	Gray, Manning & Stanwood, Boston
Webster #1 207od × 42-6 × 29-6 1727 tons 3 decks	1853 September	George Raynes, Portsmouth, N.H.	George Raynes, Portsmouth, N.H.	Spofford, Tileston & Co., New York
Eagle Wing #1 205oa × 36-6 × 23 1200 tons 2 decks	1853 October	—	James O. Curtis, Boston	Theodore Chase, Esq., Boston
Edwin Forrest #1 184od × 37 × 23-4 1074 tons 2 decks	1853 October	—	Daniel D. Kelly, East Boston	Crosby, Crocker & Co., New York

Vessel & Ref. #	*Launch*	*Designer*	*Builder*	*Original Owner*
Great Republic #29 4-mast bark 335oa × 53 × 38 4555 tons 4 decks	1853 October	Donald McKay, East Boston	Donald McKay, East Boston	Donald McKay, Boston
Romance of the Sea #2 240-9* × 39-6 × 20 1782 tons 2 decks	1853 October	George B. Upton, Boston	Donald McKay, East Boston	George B. Upton, Boston
Red Jacket #12 241-6oa × 43 × 23 2305 tons 2 decks	1853 November	Samuel H. Pook, Boston	George Thomas, Rockland, Maine	Seccomb & Taylor, Boston
Challenger #1 206od × 38-4 × 23 1400 tons 2 decks	1853 December	Samuel H. Pook, Boston	Rob't E. Jackson, East Boston	W. & F.H. Wittmore, Boston
Eringo, bark #1 113od × 26-3 × 12-1 327 tons 1 deck and half poop	1853 December	—	Brown & Lovell, East Boston	Capt. Lewin, et al., Boston
Herald of the Morning #1 202od × 37 × 24 1300 tons 2 decks	1853 December	Samuel H. Pook, Boston	Hayden & Cudworth, Medford, Mass.	Thatcher Magoun & Son, Boston
Coeur de Lion #1 198oa × 36 × 22 1100 tons 2 decks	1854 January	George Raynes, Portsmouth, N.H.	George Raynes, Portsmouth, N.H.	William F. Parrott, Boston
Lightning #1, 29 243oa × 44 × 23 2084 tons 2 decks	1854 January	Donald McKay, East Boston	Donald McKay, East Boston	James Baines & Co., Liverpool
Galatea #1 182od × 36-6 × 23 1100 tons 2 decks	1854 March	—	James Magoun, Charlestown, Mass.	William F. Weld & Co., Boston
Champion of the Seas #1 252od × 45-6 × 29 2447 tons 3 decks	1854 April	Donald McKay, East Boston	Donald McKay, East Boston	James Baines & Co., Liverpool
Sierra Nevada #2 230oa × 44-4 × 26-4 1942 tons 3 decks	1854 May	—	Tobey & Littlefield, Portsmouth, N.H.	Glidden & Williams, Boston
Lamplighter, bark #1 121* × 27-7 × 12 365 tons 1 deck	1854 June	—	Hayden & Cudworth, Medford, Mass.	Lombard & Co., et al., Boston
Barreda Brothers #1 165* × 31-6 × 21 750 tons 2 decks	1854 July	Hugh McKay, Boston	Hugh McKay, Boston	Sampson & Tappan, Boston

Vessel & Ref. #	*Launch*	*Designer*	*Builder*	*Original Owner*
James Baines #29 266oa × 44-9 × 29 2515 tons 3 decks	1854 July	Donald McKay, East Boston	Donald McKay, East Boston	James Baines & Co., Liverpool
Blue Jacket #5 235oa × 41-2 × 24 1790 tons 2 decks	1854 August	—	Rob't E. Jackson, East Boston	Seccomb & Taylor, Boston
Aurora #24, 28 200od × 43-7 × 29 2312 tons 3 decks	1854 September	Wm. H. Webb, New York	Wm. H. Webb, New York	Cornelius Grinnell, New York; Robert Grinnell, et al., Liverpool
Bounding Billow, bark #1 125od × 27-6 × 12 360 tons 1 deck	1854 September	—	Jotham Stetson, Chelsea, Mass.	Lombard Conant, et al., Boston
Chatsworth #1 190od × 36 × 24 1152 tons 2 decks	1854 September	Samuel Badger, Portsmouth, N.H.	Samuel Badger, Portsmouth N.H.	Enoch Train & Co., Boston
Commodore Perry #1 212od × 47-6 × 29 2350 tons (abt.) 3 decks	1854 September	Donald McKay, East Boston	Donald McKay, East Boston	Donald McKay, Boston
Benin, bark #1 150od × 32 × 14-8 690 tons 1 deck and hold beams	1854 October	Donald McKay, East Boston	Donald McKay, East Boston	Thomas Harrison, Liverpool
Moneynick #1 121* × 27-9 × 12-1 365 tons 1 deck	1854 October	—	Mitchell & Rice, Chelsea, Mass.	James Adger, Charleston, S.C.; Lombard & Co., et al., Boston
Saracen #1 200oa × 38-6 × 24 1300 tons 2 decks	1854 October	E. & H.O. Briggs, South Boston	E. & H.O. Briggs, South Boston	Curtis & Peabody, Boston
Charmer #1 203oa × 37 × 23 1083 tons 2 decks	1854 November	—	George W. Jackman, Jr., Newburyport, Mass.	Bush & Wildes, Boston
Japan #1 212od × 45 × 29 1964 tons 3 decks	1854 December	Donald McKay, East Boston	Donald McKay, East Boston	Donald McKay, Boston
Donald McKay #1 266od × 46 × 29 2588 tons 3 decks	1855 January	Donald McKay East Boston	Donald McKay, East Boston	James Baines & Co., Liverpool
Great Republic #8 (rebuilt) 325oa × 53 × 29-6 3356 tons 3 decks (spar deck omitted)	1855 January	N.B. Palmer, New York (rebuilding supervisor)	Sneeden & Whitlock, Greenpoint, Long Island	A.A. Low & Bro., New York

Vessel & Ref. #	*Launch*	*Designer*	*Builder*	*Original Owner*
Halifax, bark #1 112oa × 26 × 12 300 tons 1 deck	1855 January	—	C.W. Conners, Cornwallis, Nova Scotia	Clark & Jones, Halifax, Nova Scotia
War Hawk #1 193oa × 37 × 23 1074½ tons 2 decks	1855 January	—	George W. Jackman, Jr., Newburyport, Mass.	Bush & Comstock, et al., Boston
Alma #1 153 od × 34-6 × 20-6 826 tons 2 decks	1855 February	—	Ingalls & Shepard, Sullivan, Maine	Charles S. Brown, Esq., Boston
Andrew Jackson #1 222* × 41-2 × 22-3 1679 tons 2 decks	1855 March	Mason Cary Hill, Mystic, Conn.	Irons & Grinnell, Mystic, Conn.	John H. Brower & Co., New York
Lady Suffolk #1 205oa × 39 × 24 1290 tons 2 decks	1855 March	—	J.T. Foster, Medford, Mass.	Harbeck & Co., New York
Cathedral #1 227od × 40 × 30 1650 tons 3 decks	1855 April	—	Samuel Badger, Portsmouth, N.H.	Enoch Train & Co., Boston
Good Hope #1 190od × 38-6 × 24 1300 tons 2 decks	1855 April	—	James O. Curtis, Medford, Mass.	Francis Burrett, Esq., New York
Vitula #1 184od × 38-10 × 24 1196 tons 2 decks	1855 April	E. & H.O. Briggs, South Boston	E. & H.O. Briggs, South Boston	Williams & Daland
Antelope, demi-barque (barkentine) #1 155od × 27-6 × 10-6 415 tons 1 deck	1855 July	Robert B. Forbes, Boston	Ruben Burnham (in Samuel Hall's shipyard, East Boston) See *Boston Daily Atlas*, 4 August 1855	Parties in China
Emerald #1 190oa × 36-6 × 24 1120 tons 2 decks	1855 August	—	G. & T. Boole, East Boston	Howland & Ridgeway, New York
Noonday #1 197oa × 38-5 × 23-6 1177 tons 2 decks	1855 August	Fred'k Fernald, Portsmouth, N.H.	Fernald & Pettigrew, Portsmouth, N.H.	Henry Hastings, et al., Boston
Mameluke #1 212oa × 38-10 × 1156 tons 2 decks	1855 September	E. & H.O. Briggs, South Boston	E. & H.O. Briggs, South Boston	Curtis & Peabody, Boston
Elvira #1 186oa × 39-6 × 23 1150 tons 2 decks	1855 October	—	Pratt & Osgood, East Boston	Vernon Brown & Son, Boston

Vessel & Ref. #	*Launch*	*Designer*	*Builder*	*Original Owner*
Fair Wind #1 204oa × 38-10 × 24 1303 tons 2 decks	1855 October	E. & H.O. Briggs, South Boston	E. & H.O. Briggs, South Boston	Henry S. Hallett & Co., Boston
Abbott Lawrence #1 210od × 41 × 24-3 1500 tons 2 decks	1855 November	Donald McKay, East Boston	Donald McKay, East Boston	George B. Upton, Capt. Eben Caldwell, Boston
Electric Spark #1 185od × 40 × 24 1200 tons 2 decks	1855 November	—	Hayden & Cudworth, Medford, Mass.	Thatcher Magoun & Son, Boston
Golden Fleece #1 222oa × 41 × 26 1500 tons 2 decks and hold beams	1855 November	—	Paul Curtis, East Boston	William F. Weld & Co., Boston
Quickstep, bark #1 143od × 29-6 × 18-9 530 tons 2 decks	1855 November	Samuel Hall, Jr., Boston	Samuel Hall, Jr., Boston	Samuel Hall, Boston
Amos Lawrence #1 190od × 41 × 24-3 1384 tons 2 decks	1855 December	Donald McKay, East Boston	Donald McKay, East Boston	Donald McKay, Boston
Aurelia, bark #1 No data 474 tons 2 decks	1855 December	—	Hodgeman & Glover, Camden, Maine	No data. Boston
Harry Bluff #1 185od × 38-4 × 24 1166 tons 2 decks	1855 December	—	Jotham Stetson, Chelsea, Mass.	Charles R. Green, Esq., New York
Josephine, brigantine #24, 25 130od × 21-5 × 9-6 268 tons 1 deck	1855 —	Wm. H. Webb, New York	Wm. H. Webb, New York	—
Alice Tainter, bark #22, 24 140od × 31-5 × 17-8 667 tons 2 decks	1856 February	Wm. H. Webb, New York	Wm. H. Webb, New York	Post, Smith & Co., New York
Alarm #1 190oa × 38-6 × 24 1175 tons 2 decks and hold beams	1856 April	E. & H.O. Briggs, South Boston	E. & H.O. Briggs, South Boston	Baker & Morrill, Boston
Witch of the Wave #1 195oa × 37 × 24-6 1198 tons 2 decks	1856 March	—	Geo. Raynes, Jr., Portsmouth, N.H.	Unnamed parties, Newburyport, Mass.
Endeavor #1 192oa × 27 × 22 1136 tons 2 decks	1856 April	—	Rob't E. Jackson, East Boston	Cunningham Bros., Boston

Vessel & Ref. #	*Launch*	*Designer*	*Builder*	*Original Owner*
Silver Star #1 195oa × 38 × 24 1195 tons 2 decks	1856 April	—	James O. Curtis, Medford, Mass.	Reed, Wade & Co., Boston
Susan Howland #1 180od × 38-6 × 23-6 1137 tons 2 decks	1856 April	—	Pratt & Osgood, East Boston	Vernon Brown & Son, et al., Boston
Thatcher Magoun #1 200oa × 40 × 24 1248 tons 2 decks	1856 April	—	Hayden & Cudworth, Medford, Mass.	Thatcher Magoun & Son, Boston
Plutarch #1 192od × 40 × 24 1300 tons 2 decks	1856 May	—	G. & T. Boole, East Boston	Howland & Ridgeway, New York
Henry Hill, bark #1 143oa × 31 × 14-8 550 tons 1 deck and hold beams partially decked	1856 June	Donald McKay, East Boston	Donald McKay, East Boston	Charles S. Brown, Esq., Boston
Joseph Peabody #1 190oa × 38-6 × 24 1180 tons 2 decks and hold beams	1856 June	E. & H.O. Briggs, South Boston	Ewell & Jackson, East Boston	Curtis & Peabody, Boston
Flying Mist #1 200oa × 39 × 24 1184 tons 2 decks	1856 September	—	James O. Curtis, Medford, Mass.	Theodore & George B. Chase, Boston
Ocean Monarch #11 242-2pp × 45-10 × 30-2 2785 tons 2 decks	1856 September	Wm. H. Webb, New York	Wm. H. Webb, New York	Frost, Page, et al., New York
Asa Eldridge #1 204oa × 38-10 × 25 1198 tons 2 decks and hold beams	1856 October	E. & H.O. Briggs, South Boston	E. & H.O. Briggs, South Boston	Henry S. Hallett, Esq., Boston
Hesperus #1 176od × 37-5½ × 24 1050 tons 2 decks	1856 October	—	J.T. Foster, Medford, Mass.	T.B. Wales & Co., Boston
Young Turk #1 116od × 28 × 13 350 tons 1 deck	1856 November	—	James O. Curtis, Medford, Mass.	Alpheus Hardy & Co., Boston
Black Hawk #4, 24 180od × 38 × 23 1175 tons 2 decks	1856 December	Wm. H. Webb, New York	Wm. H. Webb, New York	Bucklin & Crane, New York

Vessel & Ref. #	*Launch*	*Designer*	*Builder*	*Original Owner*
Dione #1 200oa × (40 × 24 abt.) 1250 tons 2 decks	1857 March	—	G. & T. Boole, East Boston	Capt. Fales, Boston
Trieste, bark #23, 34 135 × 28-11 × 11-9 549 tons 1 deck	1857 June	Wm. H. Webb, New York	Wm. H. Webb, New York	W.A. Sale & Co., New York

Part I

How Masts and Spars Were Constructed

1

The Evolution of Masts and Spars for the Square Rigged Vessel

Masts[31]

The hull supported the masts; the masts supported the yards; the yards supported the sails; the sails filled with wind; the ship moved. That was the engine which drove the sailing ships of the 1850s.

However, many unrecorded eons and thousands of years of history have shown a slow and irresistible evolution in the development of the mast as we know it. As might be suspected, experience in sailing and an increase in the size of vessels brought on the need for developments to accommodate the latest thinking. But in the earliest of times, for which any records or artifacts have been discovered, the beginnings were quite humble.

One of the earliest records of a mast is contained in a painting of an Egyptian sailing craft that is rudely painted on a pottery vase reckoned to be from 3400 BC. This painting shows a Nile river boat with typical high prow and stern. Well forward is a short mast with a square sail. No rigging is indicated. It is generally thought that, due to the forward location of the mast and the lack of any transverse rigging, the sail was set only when the wind was almost directly from aft.

In general, the development of masts centered around the use of a fixed stick, or spar, as the mast. However, a rather strange deviation from this line of progress appeared about 2700 BC.

A mast consisting of two legs was developed. This was a bipod (two legs of a tripod). The lower ends of this structure straddled the boat's keel. The upper ends were lashed together and formed a very elongated A-frame. This entire assembly could be lowered by pivoting it about its lower end. Appropriate rigging was provided with which to accomplish this maneuver. As was previously noted this mast, also, was put into use when the wind blew from directly aft.

A detailed stone relief, from about 2700 BC, shows a small flotilla of ships of Pharaoh Sahure which was sent on a mission to harass the coast of Syria. These vessels were fitted with the bipod mast.

The relative complexity of the bipod mast apparently met with tepid reception because, except for the instances mentioned, it was never again noted in the development of masts. However, single tree masts which could be lowered appeared about 2000 BC and again in 500 BC where they were seen in Greek ships.

As time advanced the fixed mast, composed of a single tree, became the accepted form. Vessels became larger, as did their appointments, and by the year 1200 BC the new higher masts were supported fore and aft by stays and, transversely, by a form of shroud which led well aft of the mast to either gunwale.

Topping the mast was a deep, circular, basket-like structure that appeared to be of woven texture, and could conceivably hold a man to act as lookout. This was the predecessor of masthead structures.

The thousand or so years that followed showed only minor deviations from this general description, the major changes being that vessels were being built to greater dimensions.

The Mediterranean region, especially Egypt, was noteworthy for growing trees that did not yield lengths of timber that provided large, straight logs. Most trees provided lumber that was in short pieces not suitable for mastmaking.

This shortcoming was a great disadvantage in the construction of large masts. Thus, large masts fashioned from single trees were a rarity. It became necessary to make masts from groups of small timbers, fashioned to form assemblies of the required diameter. These sticks were brought together to the proper dimension and then bound tightly at intervals along their length with substantial rope lashings which, in English, became known as wooldings.

The first identified use of wooldings occurs in Roman vessels about AD 50. Several examples, in the form of reliefs in sarcophagi of the period, plainly show their existence. (Due to the depletion of suitable large trees, made-masts with rope wooldings became commonplace in the 1700s. The rope lashings gave way to iron bands in the 1800s.)

Certain Roman reliefs, dating from about 200, show the single mast with a square sail and a short extension rising above the yard. This extension supported a triangular topsail on each side of the yard. This is the earliest illustration of mast construction moving upward.

After the Roman period mast development moved from the eastern Mediterranean north and west to Europe. Spain and England were both extending their maritime horizons and this expansion fostered the development of better and larger ships.

Prior to 1000, changes were neither abrupt nor of great magnitude. The single square sail prevailed and the mast, a relatively small timber, remained as a plain spar which was now embellished with an ornamental device or a metal vane which swung according to the direction of the wind or, in many vessels, a pennant was flown.

In reconstructions based on artifacts and other findings from about 1300, the top of the mast is shown to have a utilitarian function. A structure of what would be later known as crosstrees and trestletrees supported a round, dish-like platform surrounded by protective sides. This platform, or top, was large enough to support a lookout.

For the following century or so, until the late 1400s, the elements of masting remained stagnant. The basic individual structures were the masts, many of which were made-masts with wooldings and circular tops with protective sides. Now, however, the masts, due to their increased size, were supported fore and aft by stays and, transversely, by substantial gangs of shrouds.

Toward the end of this (1400) century a new feature—one that was little noticed but of great portent—made its appearance. This was the introduction of the topmast. Initially it was a short, unimposing stick, lashed in the top, to the forward side of the protruding lower masthead. It was of length that allowed the installation of a small square sail.

(The topmast and its associated topsail would survive to become the most powerful and important square sail ever developed in both the naval and merchant services. The topsail reached proportions so large that merchant crews, with limited personnel, had difficulty in handling them. The sails became the subject of much experimentation and ultimately were divided into two smaller individual sails which prevailed for the duration of merchant sail.)

With the appearance of this new, diminutive topmast and topsail, the genie was let out of the bottle. Expansion and growth were inevitable. Ships were already being built with three permanent masts stepped in place. The size of ships did not conveniently permit the installation of an additional mast. Increasing the length of yards was impracticable due to the instability of ships and the difficulty in handling such yards. This left upward as the only option for expansion.

Shipbuilders needed no second invitation to take advantage of the option. It is stated in manuscripts of the time that topgallant masts were fitted in the English ship *Regent* as early as 1509 "in the same manner as some French vessels." The great English ship *Henry Grace a Dieu*, built in 1514, carried topgallants on her three forward masts. (She also shipped a bonaventure mizzen.)

These topmasts and topgallant masts appear to be continuations of the lower masts, resulting in spars of excessive length and difficult fabrication procedures. Also, this additional weight, located so high aloft, presented stability considerations. In extreme weather conditions it was not possible to strike these masts and relieve this topside load. A round top with raised sides was installed at the head of the topmast to accommodate the topgallant rigging.

For more than half a century these problems were dealt with and, the situation exacerbated by a general increase in the proportions of the masts and their sails, became increasingly more acute.

About 1570, possibly sooner, an innovation, attributed to the Dutch,[32] appeared. Topmasts, and at a later date topgallants, were fabricated as individual assemblies and fidded to the lower masts. Doublings at the mastheads were installed to provide continuous stiffness to the mast assemblies. Upper masts could now be rigged down in extreme weather. This innovation remained in use until the end of the square sail.

In 1600 large vessels were fitted with four masts, namely, foremast, mainmast-mizzenmast and bonaventure. The mainmast consisted of the lower, topmast, topgallant and a flagpole. The topmast and topgallant mast were fidded forward of the mast below.

The lower top remained as a round, flat dish with no raised sides. The topmast top was replaced with trestletrees and crosstrees to support the topgallant fid. The flagpole was lashed to the topgallant masthead.

Also at the beginning of this century a new mast appeared. This was the spritsail topmast. The bowsprit had long been fitted with a yard below it, which was known as the spritsail yard, and was rigged with a square sail which was called a spritsail. At an earlier date this sail was known as a blind because it completely blocked the view ahead.[32]

At the forward end of the bowsprit there was a flagstaff. This pole was replaced with a larger, more substantial mast which was called the spritsail topmast. Toward the end of the century it fell out of favor and shortly thereafter disappeared.

About 1620 the three masted ship made its appearance.[32] The fourth mast, the bonaventure, was discarded. The mizzenmast was now fitted with square sails and carried one yard less than the fore and mainmasts. From this date the three masted vessel was developed as the optimum square rigger.

In 1637, during the reign of Charles I, king of England, etc., a great ship was built to impress the maritime world, to glorify the importance of England, and to satisfy the vanity of the king. This vessel was *Sovereign of the Seas*. It was a most impressive vessel in size, armament, profligate ornamentation and, if the famous engraving by J. Payne may be used as a criterion, masting and rigging.

It was a massive assemblage of tradition in opposition to innovation and progress in contrast to experience.

She stepped three masts in accordance with the current thinking. She also retained a sprit topmast, which was losing favor. She retained the old saucer-like tops in lieu of the newer flat, timber structures. But what could arguably be her most important feature was the installation of articulated masts, i.e., those that were separate spars for each level of height, which could be down-rigged as necessity dictated.

On her fore and mainmasts, the topmasts were fidded on the forward side of the mastheads; the topgallants were fidded on the forward side of the topmasts, and so on up through the royals; the flagpoles were fidded atop the royal masts. There were trestletrees and crosstrees at each individual level.

After what seem to be the superfluities encountered in *Sovereign of the Seas* the remaining portion of the 17th century appeared to take a new turn in the name of simplification.

The most visible change was in the total elimination of the sprit topmast and the circular top from the forward end of the bowsprit.

These were replaced by a long, slender boom which lay on the bowsprit and extended about the length of the bowsprit beyond its foremost end. This extension was very clean and graceful, and provided the means of installing new sails which were called jibs. In this same time frame a new spar, the martingale boom, or dolphin striker, appeared. Its purpose was to counteract the upward forces of the jib stays when those sails were drawing.

Also, above the topmast, the articulated masts were abandoned in favor of long, slender pole masts which were composed of topgallants, royals and flagpoles in one unit. This gave the upper areas a clean, sharp appearance.

The mizzen lateen spar was sometimes retained but the portion of the sail forward of the mizzenmast was omitted. This left a sail that looked very much like a free-footed spanker. (The complete spar was retained for use as a possible replacement should one of the principal yards sustain damage.)

Tops became flat, timber structures instead of saucer-shaped assemblies and retained their circular outline on the forward edge.

The masting of vessels of American design reached ever upward. In the 1790s the keels of three frigates, *United States*, *Constitution* and *President*, were laid. The sail plan of *President* is shown with skysails. Later, in 1813, the sail plan of an American corvette, *General Pike*, shows her fitted with skysails on all masts. The truck of the mainmast was 195 feet above deck.

Skysails in American merchant vessels did not appear until the advent of the clipper ship in the 1850s. By the early 1840s all vestiges of such sails as spritsails and lateens had been relegated into the category of history. In large vessels such as packets, freighters and clippers, three masts were stepped and a bowsprit housed. General practice had proven that every advantage of sailing was to be had with this particular arrangement.[33]

In general, ships were being designed with finer, sharper lines which required the foremast to be positioned further aft. Larger hulls could accommodate greater sail area and as the sails became larger and more powerful, they also became more difficult to handle. This applied particularly to topsails, which were the largest sails in a square rigged vessel. These sails placed undue hardship and danger upon the meager crews of these ships, especially in severe weather.

This condition was so serious that, in the early quarter of the 19th century, various experimental attempts were made to ameliorate the condition. Robert Bennett Forbes, probably the foremost proponent of a solution to this problem, records in his book *Notes on Ships of the Past*,[34] that in 1824, one Daniel Tonge, Liverpool, invented a double topsail rig. At the same time, in the United States, a Capt. W. Cammett designed in 1823–24, a three masted vessel with double topsails on the fore and main masts. This vessel was called *Pan Matanzas*, and was built at Dover, New Hampshire, by R. Rogers (part owner) and W. Savage, Boston.

In 1844 Capt. Forbes himself rigged the bark *Edith* with double topsails and in 1845 rigged the ship *Massachusetts* with double topsails. In both cases the topmasts were fidded abaft the lower masts for convenience in housing when so obliged. Later, in 1847, he rigged the ship *Samoset* in the same manner except that, to please the eye, he fidded the upper masts afore the lower, in the conventional manner.

This concept of double topsails met with little and grudging acceptance over the next few years into the early years of the clipper ship. In the meantime, starting in 1850, the clipper came more and more into prominence as the desire for speed to reach California for gold became a phenomenon. With these new, speedy ships a new sail appeared—the skysail, which was rigged above the royals. This sail was confined almost exclusively to the clippers. Its life span lasted only as long as did the clipper ship.

In 1853 a man named Frederic Howes introduced an improved version of the double topsail in the clipper ship *Climax*. By this time Forbes' double topsail had gained general acceptance. However, the maritime world was quick to adopt the newer Howes version. In a few years it had become the standard rig for large square-rigged merchant vessels and remained as the standard mast arrangement until the end of the wooden sailing ship.

With Forbes' rig it was necessary to have a longer than usual masthead on the lower mast to accommodate the division of the original topsail into two new smaller sails, the lower and upper topsails. This, in turn, increased the length of the doubling of the lower mast and the topmast.

To back-fit a vessel to receive Forbes' rig was a costly undertaking since all masts had to be removed from the ship and new lower masts with longer mastheads stepped in their place. To back-fit to receive Howes' rig required no modification to the existing structure.

At this period—the 1850s—there were many other modifications to the topmast structure but none made a lasting impression on the maritime world.

Between 1853, when Frederic Howes' double topsail was introduced in the clipper *Climax*, and the painting of the mast and sail on the pottery vase from the era of the Egyptian pharaohs in 3400 BC, more than 5,000 years of progress had been made.

Bowsprit[31]

In the development of masts and spars to support the sails which propel a ship the bowsprit was a newcomer, not appearing in any form until about 1250. This date is used under the assumption that any spar independently projecting forward of a vessel's stem qualifies as a bowsprit. That definition being accepted allows this date to be used as a starting point in the evolution of this spar.

Using some 13th century English seals which depicted, by text and illustration, a town's livelihood, a vessel with a single mast and a square sail plainly indicated spars projecting forward of the vessel's bow. These were termed the beitass. They pointed forward and upward, one on the port side and one on the starboard side, and had a form of block in the end through which the bowlines from the leeches of the sail were rove.

In 1284 a seal of the English port city of Dover shows a vessel with a single mast and square sail and a single sprit projecting forward and upward through a forecastle. Except for a gang of shrouds on either side of the vessel, no rigging is indicated. Reconstruction brings the bowlines forward and through the blocks on the end of this single sprit.

Further development is indicated in 1350 on the seal of the city of Elbing. This seal depicts a ship with one mast. The stempost and the sternpost are straight timbers projecting upward from the keel at an angle of about 60 degrees from horizontal. The stempost rakes forward; the sternpost rakes aft.

There are castles at either end of the ship. The stempost penetrates the forecastle as a straight timber and continues up for a short distance, protruding like a bowsprit but at this very steep angle. Its use is still restricted to receiving the bowlines through its end.

About 1350 a small coasting boat was built and, at some time, sank in the waters surrounding Kalmar Castle in Sweden. These waters were drained between 1932 and 1934 when the castle was undergoing restoration. Among the finds that were uncovered during this project was the well preserved hull of this 1350 vessel. These remains were so complete that very little reconstruction was required.

When restored this boat, which had a hull 36 feet long, shipped a true bowsprit. Its heel was secured to the aft edge of a small platform in the bow slightly below the gunwale. Its mid length was lashed to the port side of the stemhead on the gunwale. Blocks to handle the bowlines of the square sail were attached to the bitter end of the sprit.

In 1400 a larger deep-sea vessel appears. It steps one mast with a square sail and small castles fore and aft. In the forecastle a substantial spar is housed and projects over the bow at about a 40 degree angle above horizontal. Its function is still to fairlead the bowlines through blocks secured to its end. This spar is much like the bowsprit that would soon develop.

An old drawing, circa 1470, shows a large vessel with three masts, and castles fore and aft. Protruding forward from the forecastle is a substantial spar, similar to the one noted in the previous paragraph. At its mid-length are the fairleads for the bowlines. Near the outboard end a forestay, originating in the fore top, is set up. This is the earliest example of the bowsprit functioning as a component of the standing rigging.

Toward the end of the 15th century a mural was painted in a church in Denmark. It depicts a ship with one mast and a bowsprit. On the end of the bowsprit a flagstaff is mounted vertically with a pennant flying. (This feature appears incongruous inasmuch as it does not appear again for half a century.)

About 1490 a square spritsail is slung below the bowsprit and shortly thereafter the forestay and fore topmast stay are led down to the bowsprit.

In 1545 the great (rebuilt) four masted ship *Henry Grace a Dieu* (also known as *Great Harry*) was fitted with a small sprit topmast on the end of the bowsprit. This spar was decorated with a crown at its top surmounted by a flagstaff.

By 1615 the bowsprit had become a large spar of substantial diameter and length. It now supported the forestays, the spritsail, and a new feature—a round top—set up on the bitter end of the bowsprit. The sprit topmast was now a true mast, stepped in this top and of appropriate size to support a new square sail—the sprit topsail. Above the sprit topmast a flagstaff was mounted.

In 1637 the opulent *Sovereign of the Seas* was launched and carried more sail than any other ship of the era until the ship-of-the-line *Victory* was built about 1760.

The *Sovereign of the Seas* housed a huge bowsprit which supported a round top at the end, a sprit topmast stepped in the top, and a flagstaff fidded to the sprit topmast. In addition, the bowsprit supported a large spritsail and anchored all the forestays.

This arrangement marked the zenith of the development of the bowsprit in the fashion just described.

Jibboom[31]

Early in the 18th century a new spar was introduced, namely the jibboom. This was a long, slender boom that was added to the bowsprit by doubling it inboard on its top and extending it appreciably forward. This boom received the lower end of the fore topmast stay.

A few years after the appearance of the jibboom the sprit topmast was eliminated. The sprit topsail was now slung from the mid-length of the jibboom.

In the late 1700s both the spritsail and sprit topsail were discarded. In their stead the jibboom and the flying jibboom were fitted to accommodate their appropriate sails.

At this same time, about 1790, a new spar, the martingale boom, or dolphin striker, was set up under the end of the bowsprit to aid in counteracting the upward forces generated by the headstays and sails.

This development ended the evolution of the bowsprit which remained essentially the same until the end of the days of sail.

The arrangement of the jibboom underwent a final, minor change in 1851. In that year the clipper *Invincible* was built and rigged with an additional jibboom which was called a jib-a-jib boom and was set up at the end of the outer jibboom as a separate spar. "The sail was set up as a jib topsail on the fore royal or fore topgallant stay. Its primary purpose was to overcome some of the weather helm due to the raking masts of a clipper ship."[35]

This was the final development of the jibboom in the days of the wooden square rigged merchant vessel.

Yards[35]

The evolution of yards evolved at the same time as the evolution of masts. Both are a matter of known record, as of 3400 BC, although it is an unrecorded fact that both the mast and the yard obviously existed prior to that date.

In this case, where an early vessel is painted on an early vase the yard, while not actually shown, is extending the head of a square sail.

The first actual rendering of a square yard dates from a relief carved in a stone wall about 2000 BC. This carving shows a mainmast viewed from the stern. The entire outline of the yard, while obvious and elementary, is plain and complete. A reconstruction based on the relief shows the foot of this square sail being stretched between the clews by another spar. The length of the yard appears to be about twice the beam of the vessel.

A relief in an Egyptian temple, dating from 1500 BC shows, in detail, a vessel with a single mast and a very long yard to which a square sail is bent. This yard is three times as long as the beam of the boat or more. It is made up of two lengths doubled in the slings and secured together with rope lashings.

The general scheme remained unchanged in principal until about AD 900. Two Greek manuscripts of that period have been found which show vessels with a single mast and a new spar, the lateen, to which the lateen sail was bent. The yard was in two lengths, as described before, doubled with lashings in the slings and parraled with a rope knot. The tack was brought down to deck level and the clew peaked at about 50 degrees above horizontal.

This yard, the name of which is derived from the Latin word latimus, which means sail, would progress in lockstep with the square yard until the mid–18th century at which time it disappeared.

Mosaics in Venice show that, in 1268, small craft of about 85 feet in length have two masts stepped, each rigged with lateen yards. These yards are made up of two lengths, lashed together. They are very slender and of graceful proportions.

A map of 1367 shows a crude drawing of a small vessel with a square fore yard and a main lateen. In 1400 a contemporary illustration shows a medium length yard composed of a single spar, but, in general, most vessels carried a yard made of two pieces.

Toward the end of the 15th century ships were built of greater size, hulls of more than 100 feet in length being common. Such ships required greater propulsion. This was supplied by stepping a third mast in the hulls. This mizzenmast carried a lateen yard while the fore and mainmasts crossed square yards. Illustrations indicate that these lateen yards were single sticks as were the square yards.

Until this time the main square sail increased in size as the hulls grew and was now beginning to be too difficult to handle.

A painting from 1489 shows a small vessel with a small spritsail and foresail, a large mainsail, a lateen mizzen sail and a newcomer—a diminutive square topsail above the main course. This yard was about one-third the length of the main yard and carried no rigging except a basic lift and a parral.

Ships' spar and sail plans were about to undergo long-lasting changes.

In 1492 when Columbus put out from Portugal his ship *Nina* carried a lateen sail on each of her three masts. She was a caravela latina. This rig proved to be unsuitable early in the voyage. At the Canary Islands his ships put in and *Nina* was completely remasted and rerigged with a square foresail and mainsail. The lateen mizzen was retained—she became a caravela redonda. The square rig was proving to be the rig of choice when traversing the long reaches of endless ocean.

Ships with four masts appeared in 1500. They crossed square yards on the bowsprit, foremast and mainmast with a topsail yard on the main, and lateen on the mizzen and bonaventure mizzen. This bonaventure mizzen was a new mast which became known by the new term bonaventure.

Twenty years later vessels were built with similar sparring except that topsail yards appeared on both fore and main masts. They were proportionately larger than the same yards in previous vessels.

In 1545 a picture of the rebuilt *Henry Grace a Dieu* (built originally in 1514) shows her with four masts. The fore and mainmasts crossed square lower, topsail and topgallant yards. (Topgallant yards had been seen before, as early as 1509, but this appears to have been a re-emergence of them.) The mizzenmast crossed lateen lower, topsail and topgallant yards. The bonaventure crossed a lateen lower and topsail yard.

Advancing through the years to the 17th century the lateen yard is beginning to lose favor in the larger vessels. The bonaventure has disappeared, as have lateen mizzen topsail yards. These latter have been displaced by square topsails. However, the lower mizzen lateen remains.

When this combination of sails is used an additional square mizzen lower yard must be slung above the lateen, its purpose being to spread the foot of the square mizzen topsail.

Sprit topsail yards are generally gone but the spritsail yard remains and a jibboom appears.

In 1637 the great English man-of-war *Sovereign of the Seas* was launched. Her description was made up of superlatives. She was the largest (127 feet on keel), the most potent (100 guns), the most richly decorated, the most gilded, the most highly sparred, and the most expensive vessel (£65,586) ever built up to that time.

Her sparring was extreme. The fore and mainmasts each crossed lower, topsail, topgallant and royal yards; the mizzen crossed a lateen, a square mizzen yard, a topsail yard and a topgallant yard. From her bowsprit a spritsail was slung, and her sprit topmast supported a yard for a sprit topsail. This rig appears to support the actual use of royal sails for the first time.

Up until this time, in all the artifacts, paintings, etc. that had been brought to light, there was no evidence of actual dimensions or proportions of yards. The first glimpse of such information appears with *Sovereign of the Seas.* It was observed that above the lower (course) yard the topsail yard was about half the length of the lower yard, the topgallant yard was half as long as the topsail yard, the royal was half of the topgallant and the leeches of the sails extended down in a straight line.

Early in the 1700s large three-masted ships retained the mizzen lateen but modified the sail by removing the portion of canvas that was spread forward of the mizzenmast and, shortly thereafter, the lateen was discarded in favor of a gaff. During the remainder of the century the lateen and spritsail yards disappeared. In their place came the spanker and jibsails.

About 1797 to 1800 three frigates, *Constitution, United States* and *President*, were built for the United States Navy. These were all fitted with lower, topsail, topgallant, royal and skysail yards. The skysail was unheard of before this time. In 1813 another United States naval vessel, the corvette *General Pike*, was built with a similar spar plan.

After this surge spar plans returned to more modest proportions, reaching only as high as royal yards.

This latter philosophy held sway until about 1850 when the finding of gold in California spawned the desire for faster vessels. Once again the skysail appeared and remained only as long as did the clipper ship.

The only other major change in sparring took place between 1850 and 1853 when the topsail was divided into two sections and an upper topsail yard and a lower topsail yard shared a co-existence.

Booms and Gaffs[31]

Spanker Boom and Gaff

From the middle of the 13th century the lateen yard had been a prominent component of the sparring of large vessels. It was the principal yard on both the mizzen and bonaventure masts and topmasts for at least 200 years.

Early in the 1600s lateen topsails and topgallants gave way to square sails, but the lower lateen was retained. A short time later, in larger ships, the portion of the lateen sail which stood forward of the mizzenmast was eliminated. The remaining portion of the sail became a fore-and-aft mizzen sail with its luff laced to the mizzenmast. This sail was loose footed and its head was laced to the aft portion of the lateen yard.

Toward the end of the century the lateen yard was discarded. A gaff was shipped in its stead. The foot of the new sail was extended by a long boom which overhung the vessel's taffrail. This sail was called a spanker and the boom was called the spanker boom.

In most ships the flag of the vessel's nationality was flown from the peak of the spanker gaff.

There were two types of spanker—a brailing spanker and a reefing spanker. In the former, when the sail was brailed in it was gathered in to the mizzenmast. This could be done without disturbing the spanker gaff and any flag flown therefrom. With the reefing spanker the sail was gathered down to the spanker boom. To do this required that the spanker gaff be lowered to some degree which would not be in harmony with rendering proper etiquette with the national flag.

This does not necessarily mean that all ships with monkey gaffs were fitted with reefing spankers.

A description of the monkey gaff follows that of the spencer gaff.

Spencer Gaff

The spencer appeared in about the 1840s. It was a small fore-and-aft sail which was laced to a short gaff that was fitted to the aft side of the foremast or mainmast below the top.

Most large commercial sailing vessels did not carry spencers. With those that did there was a diversity of choice. Some vessels preferred that the spencer be fitted on the foremast, some fitted it on the mainmast, and a relative few fitted the spencers to both masts. The gaff was of the length available between the masts. When used the length was usually included in the masting tables.

Monkey Gaff

This was a small gaff about 15 to 20 feet long which was fitted to the aft side of the mizzen topmast crosstrees. It was peaked at an angle and guyed to maintain a fixed position. It was used as a flag or signal gaff in many ships built in the period of the spanker.

2

Timber Used in Mast Making

In the 1850s, the forests of the Eastern Seaboard of the United States, from Maine to Georgia, provided ample timber to fulfill the demand for masts for the American merchant fleet. Among these forest trees there were five varieties of pines. Almost all the vessels listed in this book were described as being masted with one of these five varieties. By far the most popular was hard pine. For the vessels where no specific timber is recorded, one can only conjecture. Hard pine would be the most likely choice.

The vessels that differed in mast timber are *Donald McKay*, pitch pine; *Noonday*, *John Land* and *Winged Arrow*, topmasts and jibbooms, yellow pine; *Shooting Star*, topmasts and jibbooms, Southern pine, and lower yards Norway pine.

The details and characteristics of these trees are listed herewith.[38,42] All the trees are of the genus *Pinus*. The common and scientific names of each species are given, along with general characteristics, uses and natural history of the five varieties. Weight per cubic foot is taken from contemporary sources (about 1850), which vary widely. The given weights are considered to be representative dry weights. Diameters are taken at breast height (DBH) which is the official height at which such measurements are taken. Life span is the maximum age. Strength is subject to many variables but reflects a general comparison with white oak, which is given an arbitrary rate of 100.

Resistance to rot is a relative characteristic, extremely important in masting timber, and is entered herein as good, excellent and superior.

Pine, hard	*Pinus palustris*
Diameter	2½ feet, avg.; 4 feet, max.
Height	120 feet, avg.; 150 feet, max.
Life span	150 years, avg.; 300 years, max.
Weight	43 pounds per cubic foot.
Strength	90, with white oak rated at 100.

This pine, known today as longleaf pine, flourishes in the southeastern United States in coastal areas from Virginia to Texas. When specified it was always referred to as hard pine.

The trees, by their nature, grow very straight, an obvious attribute for the making of masts. The wood is very resinous, durable and straight-grained, elastic and hard.

The best hard (longleaf) pine has superior resistance to rot.

Pine, Norway	*Pinus resinosa*
Diameter	3 feet, avg.; 5 feet, max.
Height	80 feet, avg.; 120 feet, max.
Life span	350 years.
Weight	41 pounds per cubic foot.
Strength	60, with white oak rated at 100.

Norway pine is a common, misleading name for red pine. It ranges north from Pennsylvania into Canada, from the east coast of Maine westward through the Great Lakes region.

The wood is light, close-grained and rather soft. It contains a high penetration of creosote and is elastic and durable. It is always specified as Norway pine.

The best Norway pine has excellent resistance to rot.

Pine, Pitch	*Pinus echinata*
Diameter	3 feet, avg.; 4 feet, max.
Height	100 feet, avg.; 146 feet, max.
Life span	170 years, avg.; 400 years, max.
Weight	43 pounds per cubic foot.
Strength	75, with white oak rated at 100.

This species is now known as shortleaf pine or yellow pine. It is abundant throughout all the southern states from Virginia to eastern Texas. It is the tree which, in the early 1700s, was scored with the king's broad arrow and was claimed by the English crown for use in making masts for the Royal Navy.

In the 1850s the wood was widely used for large inboard timbers such as keelsons, interior planking and beams, in addition to masts. It was always specified as pitch pine and was one of the very few American woods that received high ratings in the Timber Tables of Lloyd's Register.[40]

Pitch pine retained the elasticity and durability that was essential to a good mast.

This tree should not be confused with the Northern pitch pine, *Pinus rigida*, which forms the famous Pine Barrens of New Jersey.

Pitch pine has superior resistance to rot.

Pine, Southern

This was, and still is, a common general name which could be applied to any of several varieties of pine tree indigenous to the southeastern United States. When specified it could be any one of the following species: loblolly pine, *Pinus taeda*; longleaf pine, *Pinus palustris*; or shortleaf pine, *Pinus echinata*. All of these have the characteristics required of a good mast.

Southern pine has superior resistance to rot.

Pine, Yellow

Like Southern pine, yellow pine is a common general name that could be applied to any of several species of pine tree indigenous to the southeastern United States. When specified it could be any one of the following species: loblolly pine, *Pinus taeda*; longleaf pine, *Pinus palustris*; or shortleaf pine, *Pinus echinata*.

Yellow pine has superior resistance to rot.

In the List of Known Mast and Spar Dimensions included in this book there are many vessels for which no known wood has been specified. It appears reasonable to assume that all these vessels had masting of American pine. However, at this period (1850s), most vessels sailed for faraway ports with a full burden and at their destination, after off-loading their cargo, loaded many items that could be disposed of at a profit back home. Among such articles there is the possibility that two types of select masting timber might be brought to the United States for use by domestic shipbuilders. These woods were poon, from the East Indies and Ceylon, and cowdie (kauri) from New Zealand.[39,41]

Poon was used for masting ships in India, while cowdie was generally brought back to England. Both woods were held in high esteem for mast making. As a matter of general information their physical statistics (as complete as are known) follow.

Poon	*Calophyllum tomentosum* (*Calophyllum* = beautiful leaf; tomentosum-hairy)
Diameter	4 feet, max.
Height	150 feet, max.
Life span	
Weight	35 pounds per cubic foot.
Strength	Has fine strength for its weight.

At first glance the wood appears similar to mahogany with darker wavy streaks of reddish brown lines. With care the wood may be properly seasoned despite its tendency to warp. It is much sought after in the East for the masts and spars of boats.

Cowdie	(kauri, New Zealand), *Agathis australis*
Diameter	3 feet avg.; 10 feet max.
Height	90–100 feet, clear of branches.
Life span	
Weight	30–39 pounds per cubic foot.
Strength	More durable than pine.

Like pine, cowdie is coniferous and contains a considerable quantity of resin which exudes from it spontaneously. It has lightness and elasticity which makes it desirable for the masts of frigates and three-deckers.

In full-grown trees the girth breast high (DBH) is maintained for a height of 60 feet above the ground.

Any timber used for masts must possess strength and durability in addition to lightness, flexibility and elasticity. If the mast has rigidity that will not yield to the sudden impulses to which it is subject, it soon becomes fractured; if it lacks elasticity the fiber becomes upset and the mast becomes weaker.

The most obvious requirement for a tree to be suitable for masting is that it be of straight growth. In densely covered forests, where the sunlight is sorely restricted, the trees reach for the light and their trunks grow straight as an arrow for their entire height.

John Fincham[39] lists the following qualities which trees for masts and yards should have.

1st The wood should be of a clear yellow, and not white; as this color indicates that it is more resinous.

2nd The grain should be fine and close, the heaviest wood of each species being the most esteemed.

3rd The concentric circles in the body of the tree should not be too thick; they should be alternately of a brilliant and dark yellow, and full of resin.

4th When stripped of its bark, a resinous juice of a pleasant smell ought to exude from all parts, if it remain exposed to the sun.

5th If the tree is of a dark red, and the resinous juice is of a blackish color, it shows incipient decay.

6th The fir should have attained a proper age before it can be considered in perfection; young trees have too much sap.

7th The color of the wood must be uniform; if, when cut, it shows veins or varieties of colors, it is not good.

8th It should have neither shakes nor rind-galls, nor too great a number of knots; attention must likewise be paid, to see that it is not rotten. To hide rottenness, the merchants will make a knot appear very sound with hot resin, which is difficult to discover.

(A shake is a natural split in timber; rind-gall is damage a tree receives when young, so that the bark or rind grows in the inner substance of the tree.)

3

Harvesting, Storage and Preservation of Timber

On the day the white man first stepped ashore in the New World, the axe came ashore with him. In an unknown and inhospitable land this was his indispensable tool. It would provide him with wood for his fires, logs for his shelter and protection from his enemies, whether animal or man.

It was an age when wood was the predominant commodity and a necessary ingredient for every being's survival.

After individual needs had been fulfilled community needs required attention. In almost any direction wood was a necessity. It is probably safe to state that logging was the first large scale activity to be established in the new country.

Logs of all sizes, shapes and species of wood were required for shipbuilding, public structures and habitation. The source of most of the required timber lay at the population's doorstep.

The entire East Coast of North America, from Georgia to Maine, was forested with an abundance of pines, oaks, firs, walnut, chestnut and other species that were there for the taking.

Of all the activities that required timber none was more voracious than shipbuilding. And none required the variety of unique shapes and huge sizes than did this industry. The trees that were most vulnerable to the axe were the majestic, first growth giants that had been saplings long before the arrival of Christopher Columbus in the New World.

Axemen, however, gave no quarter and when they ventured into the forests trees of all dimensions and varieties fell before them to meet the demands for timber.

Harvesting trees for making masts was an arduous and expensive undertaking. The first task was to assemble a gang of expert axemen[43] and associated workers. Needed were sawyers to man the huge cross-cut saws that would cut partway through the trunk and determine the direction in which the tree would fall. These men also cut the felled trees into desired lengths.

Also needed were choppers, the men who were expert with the narrow axe which was actually used to fell the tree. There were also hewers, men who were adept at squaring timbers to lines scored in the raw log to given dimensions. This operation was done with a broadaxe.

Trees for masts were felled and converted to four-squared sticks right on the spot. Hewing them to the largest square that the fallen trunk could provide was economical in several ways. The removal of excess weight made transporting the trimmed stick less laborious and also reduced the amount of material that would have to be removed in the mast yard to bring the stick down to its required four-square diameter.

Lowest on the list were the ordinaries. These were apprentices or novices who were learning the skills of handling an axe. Their task was to clear brush and young growth to allow access roads to be formed over which to transport felled timber, and to perform general clean-up work.

A gang was supported by a considerable amount of equipment and hardware to set up necessary living requirements in a camp: cooking and food supplies; carts for hauling, and miscellaneous provisions; provender and bedding for the dray animals; and big wheels for transporting felled trees to the nearest landing or water.

In selecting a given tree for a mast the first and most obvious requirement was that it be straight. The second requirement was that it be of usable length. Generally, agents who were experts at evaluating trees by external inspection decided which trees should be felled for masts. Some rind-galls, which are imperfections that have penetrated beneath the bark into the wood fiber, can render a tree unsuitable for mastmaking. The same goes for any visible cracks caused by stresses or unusual weather conditions.

One defect that cannot be detected by even the best agent is internal decay or rot. Of all the trees that are felled a surprisingly high percentage are afflicted to some degree with decay in some portion of the trunk. Trees in this condition are applied to a different use as their condition warrants.

All trees, when felled, were trimmed clean of branches and foliage for their entire length. Trimming eliminated excess weight when being hauled and excessive space when stored. In many cases trimmed trees were squared-off so that they became sticks. Often this detail was spelled out in a timber contract.

If trees were felled and, for some reason, could not be hauled from the site in the near future they were completely covered with brush to protect them from sunlight and air which would quickly dry them out, thus inducing warping and splitting.

The number of pine trees from which the largest masts were made from a single trunk or stick was numerous but limited. At a time before the American Revolution, when the king's agents combed America's forests for the largest and best trees from which to make masts, the Hudson River agent reported that he could supply trees to make masts, 30 inches in diameter but that 32 inch trees would be difficult to find.[15]

Since that time almost all masts of about 28 inches diameter or more are made-masts, which are reminiscent of a collection of faggots bundled together with wrappings.

In actuality the made-mast is considered superior to the single stick for two reasons, namely, that the bundle of sticks, tightly bound, is inherently stronger than its equivalent in a single timber, and also the individual sticks in the bundle may be inspected and separately accepted or rejected. In a single stick mast there is always the possibility that incipient decay lies hidden from sight.

Once a tree had been felled the next problem was to transport it to the populated centers where it was needed. The solution to this problem was the building of logging roads from the logging camps to the nearest waterways where the timber could be floated to its destination at a river or the seacoast.

The most important and efficient method of transport were the big wheels also known as a logging cart or even merely a cart. These carts consisted of a pair of wheels and an axle. The wheels were about eight to 12 feet in diameter depending upon the size of log that might be encountered. The reason for this great diameter was to achieve a working height above the ground. This height was about equal to the radius of the wheel. Clearance under the axle was the working space.

The wheels were made up of a wooden hub, heavy wooden spokes and rim, and a flat iron tire six or more inches wide. The wheels, when assembled on the axle had a gauge—the distance between wheels—of about six feet.

When a log was ready to be transported the wheels were positioned above it so that they straddled it with the log centered below.

With short, heavy logs the cart was placed with the axle at the midpoint or balance point of the log. A chain sling was passed over the axle and then down under the log, which was levered up so as to be clear of the ground, and the chain brought up again to the axle. This was a considerable task with a log, or stick, that could possibly weigh a ton. When the chain was secure the log could be transported in a horizontal position, totally clear of the ground.

This method was the fastest way of hauling and also had the decided advantage of protecting the log from injury through contact with the rough roadway.

In some cases, where smaller sticks were involved, these carts could be moved by manpower. However, with large logs the motive power was produced by teams of horses or oxen, especially where considerable distances were to be covered.

The general philosophy of having the load underslung was that it was much less work to pass the axle over the log than it was to raise the log over the axle.

The above description of this operation is not complete in detail. Rather, it describes the general intent of the procedure. A complete description may be found in Ref.[43]

There was a more acceptable method of moving timber for masts, especially for sticks which could exceed 100 feet in length. This method was to skid the trees. In this process the cart was placed over the butt end of the felled tree and the chain sling passed over the axle, down under the tree, and up again to the axle. This end was then lifted clear of the ground. The top end of the tree was then left to drag along the road. Horses or oxen were used to haul this load to the landing, cove or pond which had been built or dug as a temporary storage facility.

Snaking was a third method of moving felled trees and was usually employed where there were no logging roads and distances were short. In such cases the log was hitched directly to the team of horses or oxen and dragged, butt end first, along the ground, the log carving its own path as it was hauled to its destination.

In the cases of masting timber, where external blemishes to the bark, such as rind-gall and surface fractures, were so much abhorred, this does not seem to have been a method of choice.

As may be concluded from the above descriptions the handling of masting timber was a prodigious task. However, even with that being the case, the procurement of adequate mast material did not place undue hardship on the timbermen in the United States. Usually the stands of timber were of relatively close and easy access.

To illustrate the potential magnitude of the problem of bringing mast timbers to the shipyard the following text is quoted from French archives and repeated by Fincham[39] in his book on masting.

> In 1765, the King of France (Louis XVI), to lessen his dependence on the northern states for the supplies of the different firs, sent proper officers to undertake the conveying of timber from the Pyrenees to the Depot at Bayonne; the principal direction of this great undertaking was entrusted to M. LeRoy, an Engineer of the Royal Dockyards; and, according to his own memorial, he appears, in prosecuting this vast undertaking, to have cut through rocks in many parts to the extent of four or 5,000 toises, to get at the proper places for forming the declivity at the height of nearly 1800 feet above the plains. In

some places bridges were thrown over ravines 50 toises broad; and different causeways formed, with a proper inclination, and sufficiently wide to turn the longest tree at the different windings.

A toise, pronounced "toiz," was an old French measurement of length equaling 1.949 meters, or 6.395 feet.

No such formidable obstacles confronted timbermen along the East Coast of the United States.

All trees, once they were felled, required seasoning, and the predominant method was to totally immerse the logs in water. This was the reason that coves and ponds were dug so that the timber could be preserved while awaiting shipment to a final destination. If kept in this state for many months the timber might be ready for immediate use at its destination.

For centuries prior to the 1850s it had been common knowledge among shipbuilders and others who dealt in the processing of trees that wood required seasoning before it could become stable and ready to be transformed into usable forms. Seasoning, like many other words, had a different interpretation among different people.

To some woodsmen and woodworkers in early America seasoned meant that the wood had been cut in the proper season of the year, during the winter when the sap was down. However, this interpretation was in the minority. To most woodsmen a log was seasoned after it had spent a long time curing, especially under water.

Many proverbs originated, some being myth and others hiding fact. None applied directly to shipbuilding but are, nevertheless, of interest.[44] One almanac states, "If you would have your flat boards lay, hewe them out in March or Maye." Another advises, "When the moon is new to full, timber fibers warp and pull." Still another proverb, which may conceivably be factual, says, "Wood that lives with face away from sun, doesn't warp when sap has run." This suggests that wood cut from the north side of a tree (where green moss is prevalent and where the sun shines the least) is less liable to warp from further dampness when worked.

While these sayings may appear to be nothing more than proverbs it should be remembered that they all originated from men's observations in their relationship with wood.

An early description regarding seasoning appeared in *The Builder's Handbook*, 1732, and is as follows, in part.

> There are some who keep their timber submerged in water to hinder the cleaving (splitting); this is good in fir (pine), both for better stripping and seasoning; yea, not only in fir but in other timber as well. Lay, therefore, your boards in water (if running as in some mill stream, the better).... Elm felled ever so green, for sudden use, if plunged for five days in salt water obtains admirable seasoning and may be immediately used. Water seasoning is not only a remedy against the worm but for its efficacy against warping and distortion of timber. There are some who bury their timber in the earth, others in wheat. There are those who season by heating with fire. But for general seasoning and quick use, the soaking methods seem best.[44]

Moving logs down waterways is essential in many densely forested regions. In log form the wood is not damaged by water immersion. In fact, water curbs insect and fungus attacks, prevents rotting and greatly reduces splitting.[41] The use of unseasoned timber was a principal cause of dry-rot spores spreading throughout a wooden vessel. Seasoning removed the sap, and proper air seasoning had to be slow in order to dry the wood evenly, thus making it more durable. Drying the wood too rapidly causes it to twist and check; therefore it was desirable to fell trees in winter when the sap was down. If shipments to shipyards were delayed, timber was often submerged in water until it could be removed for subsequent air-drying.[43]

For a discussion of wood preservation as it applies specifically to ship and mast timbers the following is taken from John Fincham's *Treatise on Masting*, 1854.[39] This book was intended to apply specifically to the British Navy but its implications are contemporary and universal.

> To preserve a store of mast timber, various methods have been used. Most kinds of fir [pine] timber for masting require the same means to be used for their preservation; but timbers of different qualities require different degrees of attention. Timber that is hard and fine grained, or that contains a great quantity of resin, will preserve its strength longer in dry sheds than when it is exposed to the vicissitudes of weather. On the contrary, timber which contains little resin ... requires to be constantly under water or sheltered from the influence of variable atmospheres. For in this kind of timber constant changes of temperature produce more readily, and in greater degree than any other, the alternate motion of the fibre, which not only assists the evaporation of those particles, which constitute the adhesiveness of its fibre, and the essential principle of its strength, but produce fermentation and destroy its durability.
>
> The means generally used to preserve timber for masting, is to submerge the sticks under water in locks, or bury them in the sand or mud of the seashore, though it is not uncommon to let them float and be occasionally turned. This practice was not condoned due to the constantly distending and contracting of its fibres, which greatly impaired a log's strength. When the timber is totally immersed in mud or water, it is exposed to the least variable element, and, consequently subjected to the least change.
>
> To bury fir [pine] timber in mud appears, from long experience, to be best suited for its preservation; but with large and light sticks, as yellow pine bowsprits, etc., there is some difficulty in keeping them down by the weight of mud, especially after it has once been disturbed; for which reason locks are constructed between brick or stone piers, to keep them constantly under water, and to contain several heights, so as to have them separate, according to their specifications.
>
> It is reported that, in France, at Brest it was the routine practice to bury all fir timber for mast purposes in mud; and that, when the weight of the mud was not sufficient to keep

them down, piles were driven in rows on each side of different tiers of sticks, and pieces fixed across them to prevent their rising; that, at Toulon, they were kept down by punts filled with stone, and sunk upon them; and that, at Rochfort, channels or moats were dug just below high-water mark at spring tides, and piles driven on each side to take three masts abreast between them, with cross pieces or chains attached to them to keep them down. Since it appears, as well from the experience in our own [British] dockyards, as from the testimony of the French and other nations, that the best means to preserve a store of mast timber, is to bury it in mud, or to keep it wholly under water, the mode of executing it can present but little difficulty, as various may be the plans for performing it; but it will appear that it must depend, in a great measure, on circumstances connected with the place where the timber is to be preserved.[39]

Very little space in American literature is devoted to the subject of the preservation of timber. References 41, 43 and 44 touch on the subject in a non-nautical manner, all being interested in wood in general.

However, from the information that is available, we know that all sources, in any time frame, agree that the way to preserve felled timber is to submerge it in water.

Correspondence between shipbuilder William H. Webb, of New York, and several business personages gives an insight into the method of storing and preserving timber along the East Coast of the United States.

It is an established fact that trees, after being felled, were hauled from the forest to some source of water such as large creeks, rivers or to the seashore itself. These destinations were invariably the sites of huge ponds, coves or man-made lakes. They also were located as closely as possible to shipbuilding yards.

These facilities were of considerable size, correspondence implying that they could hold thousands of logs measuring as much as 100 feet in length and 36 or more inches in diameter.

In these holding ponds timber could remain seasoning for a year or two if the trees had been harvested on speculation. On the other hand, trees cut under contract could be shipped out in a short time. However, the principal feature of the pond was that it provided a convenient place to store timber while it was seasoning, a process measured in months.

We know that all timber was tagged, measured for length, breadth and depth (if previously squared), and quantified in cubic feet per log. The price of a given log was based on the overall size of the timber but was determined by volume in cubic feet.

The individual logs were also identified by species of tree. Similar varieties were stored together.

The Webb correspondence reveals that there were at least three timber holding facilities in the area of New York City. These were, namely, Fitch's Basin, Jersey City; Menzies' Basin, New York City; and Brown's Basin, Hoboken, New

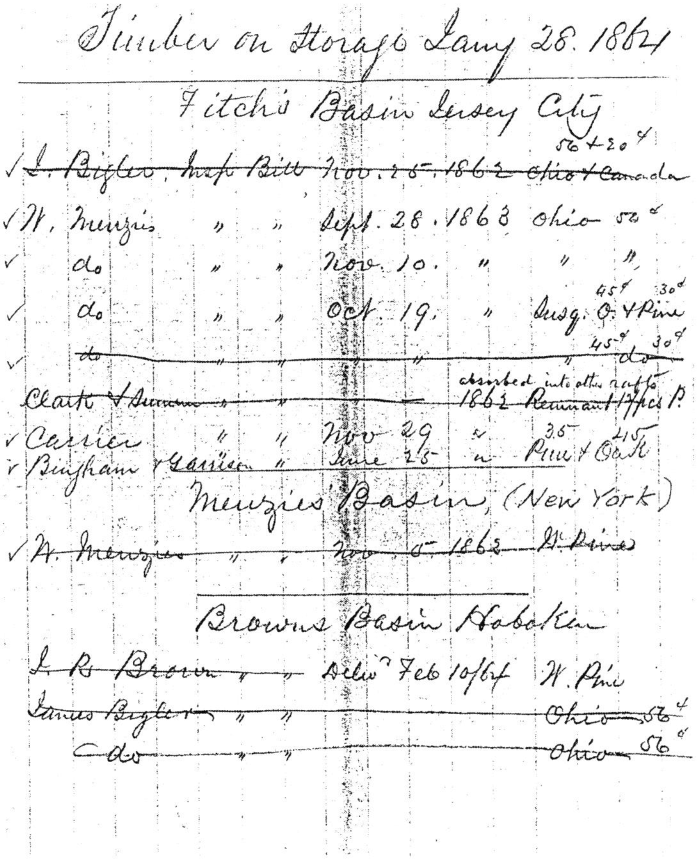
Timber on Storage Jany 28. 1864
Fitch's Basin Jersey City
Menzies' Basin (New York)
Browns Basin Hoboken

Figure 3.1. Timber in storage, January 28, 1864.

Jersey. The correspondence also indicates that there was a holding pond on the Hudson River, at Troy, New York.

A short, miscellaneous note, among the correspondence, confirms the existence of these facilities and is reproduced in Figure 3.1.

This proximity of holding ponds to shipyards and forested areas was probably a repeated scene along the eastern seaboard of the United States.

Every shipyard maintained a ready service inventory of small stuff, such as knees, crotches, stanchions and the like, that were required in great numbers but individually did not occupy much space. These were used in construction at a fast pace and were not overly difficult to handle.

Large timbers, such as masts and spars, could range from 30 to 100 feet long. In a busy shipyard it is conceivable that extremely long sticks might be subject to inadvertent damage. This could not be tolerated with masting logs, which were desired to be free of all blemishes if possible. Such timbers might be expected to season for months, hence, the efficacy of the holding ponds or basins for storing and seasoning timber safely. From these safe havens the various shipyards could order out their needed timber at will.

As stated previously each stick was measured and tagged and, in the case of William H. Webb, each individual stick was marked "W H W" in black paint.

Parcels of timber for various owners were stored together in separate groups called lockings. A locking could consist of approximately 20 or fewer sticks. When large shipments were to be moved several lockings (as many as six) were assembled together to form a raft. The raft was then floated to its destination.

In addition to individual tagging of sticks some groups of timber, possibly of superior quality, were given specific titles. At one time, in 1863, Fitch's Basin was holding one group of timber which was designated "Our Own Ohio Oak Timber," and another designated as the "Snow Shoe Timber." The signification of the terms is not known.

The foregoing discourse on the storage and preservation of ship timber pertained to all such wood, not only to material for ships' masts. However, it is a subject which receives little attention in any works that are devoted to ship design and construction. Nevertheless, it is an important ingredient in the completion of the final product and, within its confines, is of some limited interest and importance. With this in mind the following pieces of correspondence between the holding basins and William H. Webb (the company) are included in order to complete the picture.

The transfer of material from one facility to another was a straightforward and uncomplicated procedure as is attested by the following communications.

On April 22, 1863, A. J. Fitch sold and agreed to deliver 25,000 cubic feet of ship timber, known as "Our Own Ohio Oak," to William H. Webb. The timber was to average 80 cubic feet per stick. The order was to be processed as soon as possible during the present season. The timber would be delivered by way of Troy, New York, at which location it would be decided whether to ship it to Fitch's Basin in Jersey City or directly to Webb's shipyard. The price was 50 cents per cubic foot.

The order was delivered on September 28, 1863. The total delivery was 20,687 cubic feet, which is annotated as an addition to the bill of sale. Also identified were the number of sticks of each type of wood. A separate note indicated that the timber was delivered to Fitch's Basin in Jersey City.

Figure 3.2 is a copy of the above bill of sale.

On September 19, 1863, William Menzies sent a directive to A. J. Fitch, in Jersey City, to deliver to William H. Webb two tons (262 sticks) of Ohio Oak Timber.

This timber was received by Menzies at his basin in New York to be held and then delivered to Webb as needed. On September 29, 1863, Fitch's receipt acknowledged the 262 sticks and also itemized them by species of wood and quantity of each.

A copy of each piece of correspondence is included here, as Figures 3.3 and 3.4.

In both letters the species of wood are noted. All are identified except one which is noted as "W Wood" (whitewood).

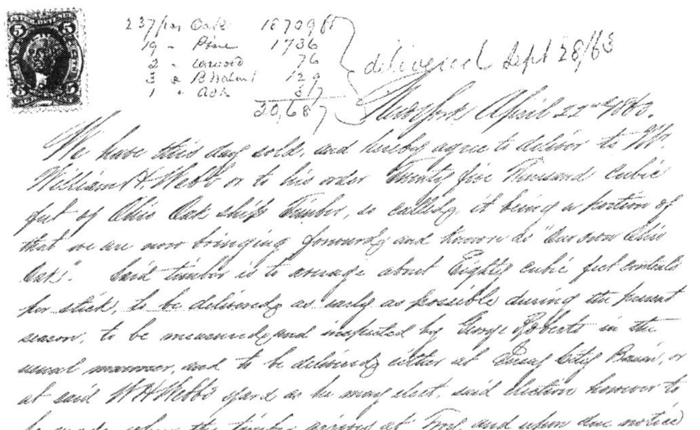

New York April 22 1863.

We have this day sold, and hereby agree to deliver to Mr. William H. Webb on to his order Twenty five Thousand cubic feet of Ohio Oak ship Timber, so called, it being a portion of that we are now bringing forward and known as "Our own Ohio Oak." Said timber is to average about Eighty cubic feet contents per stick, to be delivered as early as possible during the present season, to be measured and inspected by George Roberts in the usual manner, and to be delivered either at Jersey City Basin, or at said WH Webb's yard as he may elect, said election however to be made when the timber arrives at Troy and upon due notice thereof to him by us, and is to be settled and paid for at the rate and price of Fifty cents per cubic foot, cash on delivery of survey bill, seconds half price, and each to pay one half of the charges and expenses of the survey.

Menzies Vicke & Mather

It is understood that Menzies Vicke & Mather are to deliver this timber as above stated, and that the measurement and inspection will be made after the arrival of the timber at Jersey City Basin or My Ship Yard.

Menzies Vicke & Mather

Figure 3.2. Bill of Sale for 25,000 cubic feet of timber, 4-22-1863. Transcription at right.

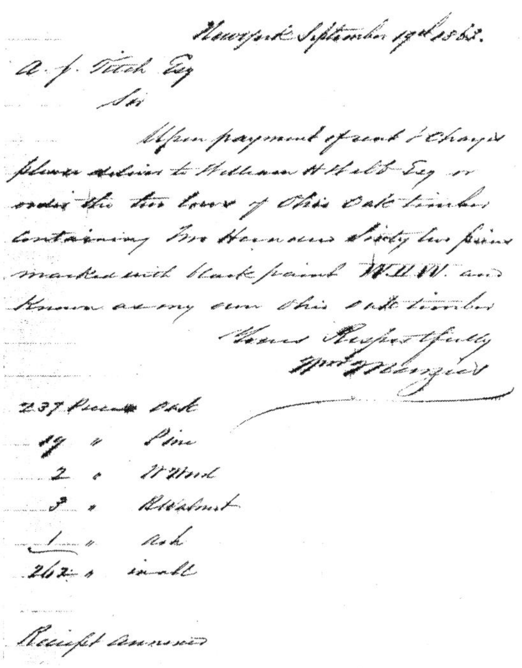

New York, September 19th 1863.

A.J. Fitch, Esq.
Sir

Upon payment of rent & charges please deliver to William H. Webb, Esq on order the two tons of Ohio Oak timber containing Two Hundred Sixty two pieces marked with black paint W.H.W. and known as my own Ohio Oak timber.

Yours respectfully
Wm Menzies

237 pieces Oak
19 " Pine
2 " W Wood
3 " B Walnut
1 " Ash
262 " in all

Receipt annexed

Figure 3.3. Correspondence from Wm. Menzies to A. J. Fitch, Esq., re delivery of timber, 9-19-1863. Transcription at right.

Whitewood could be American basswood (*Tilia Americana*), also known as American linden and whitewood; or Yellow poplar (*liriodendron tulipifera*) also known as poplar, tulip tree and whitewood.[41]

As previously stated timber to be moved from one storage facility to another was gathered into groupings known as lockings and rafts, a raft being a group of six (maximum) lockings. Movement of these groups was generated by simple letters or correspondence requesting the action to be taken. Paperwork was reduced to a minimum. Some examples are noted below.

On November 14, 1863, a letter from A. J. Fitch, Jersey City, acknowledges receipt of four rafts, made up of 356 sticks of oak and pine, from William Menzies for storage in Fitch's basin. The wood had William H. Webb's "W H W" initials painted on each stick and was known as the "Snow Shoe Timber."

On November 23, 1863, a brief directive from William Menzies to A. J. Fitch ordered delivery of the Snow Shoe Timber, after payment of rent and charges, to William H. Webb. These directives are included herein as Figure 3.5.

An order dated November 17, 1864, from William Menzies to A. J. Fitch, authorizes the delivery of four lockings of Ohio Oak timber to William H. Webb. All sticks were marked "W H W" in black paint. The order is included as Figure 3.6.

The logistics involved in converting a full grown tree in the forest into a shaped timber in a ship's hull are complicated and of great physical magnitude. In general, our thoughts are confined to the activities within the shipyard. How the

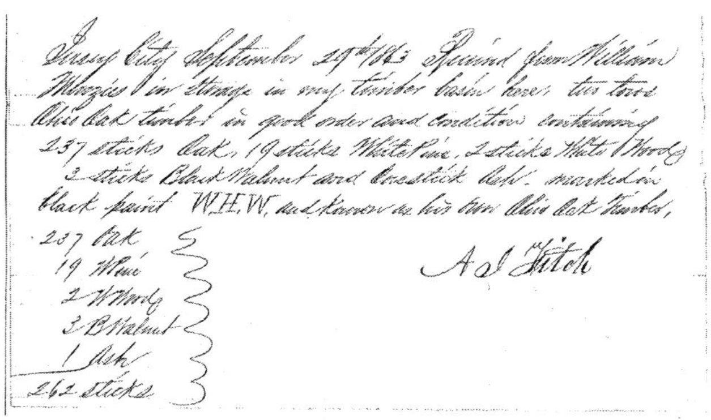

Jersey City September 29th 1863. Received from William Menzies in storage in my timber basin here, two tons Ohio Oak timber in good order and condition containing 237 sticks Oak, 19 sticks White Pine, 2 sticks White Wood, 3 sticks Black Walnut and One stick Ash, marked in black paint W.H.W. and known as his own Ohio Oak Timber.

A. J. Fitch.

237 Oak
19 Pine
2 W Wood
3 B Walnut
1 Ash
262 sticks

Figure 3.4. Correspondence from A.J. Fitch, Esq., to Wm. Menzies, re receipt of timber, 9-29-1863. Transcription at right.

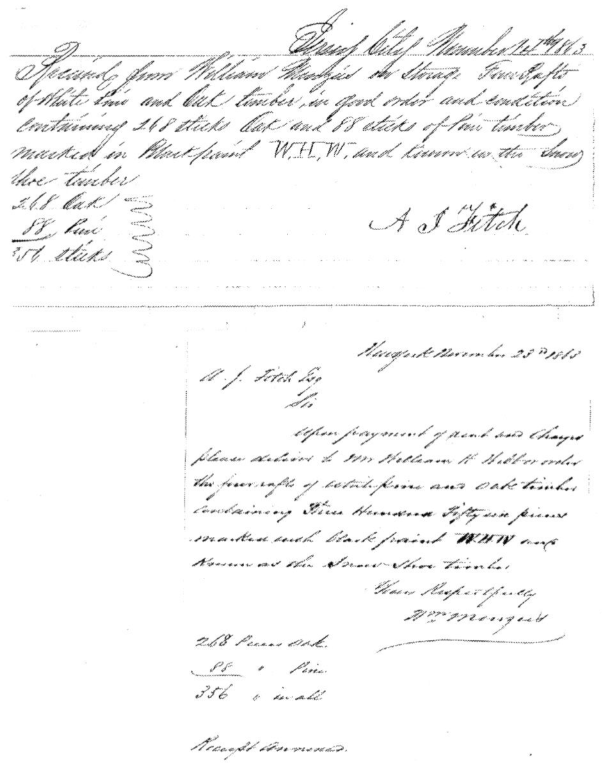

Jersey City November 14th 1863

Received from William Menzies in Storage Four Rafts of White Pine and Oak timber in good order and condition containing 268 sticks Oak and 88 sticks of Pine timber marked in Black paint W.H.W. and known as the Snow Shoe timber.

268 Oak
88 Pine
356 sticks

A. J. Fitch.

New York, November 23rd, 1863

A. J. Fitch Esq.

Sir:

Upon payment of rent and charges please deliver to Mr. William H. Webb on order the four rafts of White Pine and Oak timber containing Three Hundred Fifty six pieces marked with black paint W.H.W. and known as the Snow Shoe timber.

Yours respectfully,
Wm Menzies

268 pieces Oak
88 " Pine
356 " in all

Receipt annexed.

Figure 3.5. Letters regarding movement of "rafts" of "Snow Shoe Timber," November 14–23, 1863. Transciption at right.

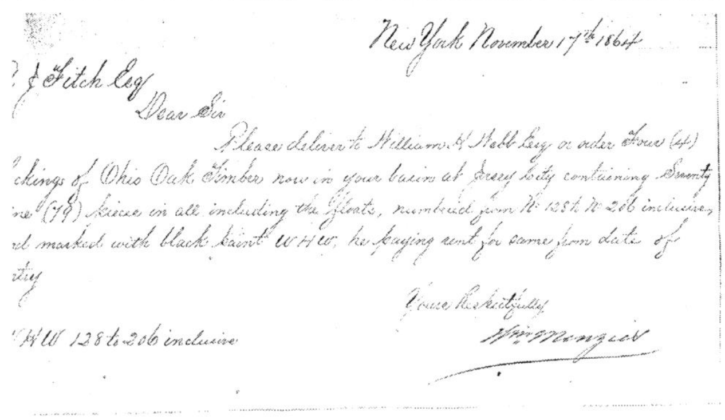

New York, November 17th 1864

A.J. Fitch Esq.

Dear Sir

Please deliver to William H. Webb Esq on order Four (4) Lockings of Ohio Oak Timber now in your basin at Jersey City containing Seventy nine (79) pieces in all including the floats, numbered from No 128 to No 206 inclusive, and marked with black paint WHW he paying rent for same from date of entry.

Yours respectfully,
Wm Menzies

WHW 128 to 206 inclusive

Figure 3.6. An order to deliver four "lockings" of timber, 11-17-1864. Transcription at right.

various materials arrive on the building site is of little concern and interest but, nevertheless, is a vital part of the process.

The size and weight of many ship timbers rendered it impracticable for transporting on land. Roadways and wheeled vehicles did not lend themselves to the task. In addition, shipyards, by their nature, provided direct access to waterborne traffic.

A final positive feature of transportation by water was that constant immersion of the timber contributed directly to the required preservation of the wood.

4

Proportions of Masts and Spars

Proportion controls the size of masts in a ship and is governed by two reference categories. The first category involves the length of the vessel and the fullness of its body. The second category is set up by knowledgeable craftsmen, the mastmakers who gathered information over time and established empirical formulae and details which were then applied to the making of masts. Details and proportions were adopted which combined strength, utility and appearance. The criteria were accepted and masts were made in accordance.

Among the first problems requiring solution were those of determining the most advantageous number of masts that could be stepped, and the maximum height that a mast could be without endangering the stability of a ship.

In reference to the first problem Kipping[33] writes, "The masts being used to extend the sails by means of their yards, it is evident that if their number were multiplied beyond what is necessary, the yards must be extremely short, that they may not entangle each other in working the ship, and by consequence their sails will be very narrow, and receive a small portion of the wind. If, on the contrary, there is not a sufficient number of masts in the vessel, the yards will be too large and heavy, and cannot be managed without difficulty. There is a mean between these extremes which experience and the general practice of the sea have determined, by which it appears that in large ships every advantage of sailing is retained by three masts and a bowsprit."

The number of masts in smaller vessels was, in general, determined by the size of the vessel.

Concerning the height of masts Kipping[33] continues,

> The exact height of the masts in proportion to the size and form of the ship remains yet a problem to be determined. The more the masts are elevated above the center of gravity the greater will be the surface of the sail which they are enabled to present to the wind,—so far an additional height seems to be advantageous. But this advantage is diminished by the vibrating movement of the mast, which operates to make the vessel stoop to its effort; and this inclination is increased in proportion to the additional height of the masts—an inconvenience which it is necessary to guard against: thus what is gained on one hand is lost upon the other. To reconcile these differences, it is certain that the height of the mast ought to be determined by the inclination of the vessel, and that the point of her greatest inclination should be the term of the height above the center of gravity.
>
> With regard to the general practice of determining the height of the masts, the extreme breadth of the ship from out to out has been admitted by long use the best rule for determining the length of the masts, that they may have proper support by the spread of the rigging.[33]

From the above information it would appear that, once the height of the mainmast was determined, the foremast and the mizzen mast would be of the same height as the mainmast and all three would be rigged with yards of uniform length, which when combined presented the maximum surface of sail to the wind.

This arrangement was not acceptable because it brought the center of effort of the sails to a height which tended to incline the vessel to a degree that jeopardized her stability.

In addition, it was found that, with this cloud of sail, the mizzen sails inhibited the ability of the sails on the mainmast and foremast to perform with their maximum efficiency.

Finally, the great surface of sail on the mizzenmast significantly influenced the handling of the vessel.

Fincham[39] injects the importance of the sizes and proportions of the sails themselves, being carefully selected in relation to each other, as the depth of the topsails to that of the courses, of topgallant sails to topsails or the proportions of the sails on the fore and mizzenmasts, to those on the mainmast. It has been observed that when the proportions had been carried to extremes, experience had soon discovered the error, and determined the limits. Thus, making the sails too nearly equal, on the main and foremasts, cannot be done without pressing the ship too much forward, or by carrying the foremast back, which not only obstructs the proper working of the yards, but prevents the wind from having its full effect on all the sails. If the topsails are too deep, in relation to the courses, there is difficulty in shifting a topmast; if the topgallants are too square, it is too difficult to support the masts above, or if too narrow, the proper area of sail cannot be obtained without making the sail too taunt.[39]

General experience, rather than plans and theory, has governed the sizes of sparring of merchant ships. Notably the diameters of masts and yards have been determined by observation and the history of similar ships.

Remedies for these situations were sought and found through formula and experiment. To reduce the height of the center of effort a formula was adopted that limited the height of the mainmast, and the yards above the topsail yard were successively reduced in length.

The diameters and configurations of masts and spars were standardized. The largest or given diameters were taken at the following locations on the spars. In lower or standing masts at the partners; topmasts and topgallant masts at the caps or to the mastheads below; bowsprits at the bed; square yards at the slings or mid-point; in jibbooms at the bowsprit cap; in driver or spanker booms, one-third from the after end, or at the taffrail; studdingsail booms, topsail and above, one-third from each end, and parallel between; lower studdingsail booms, one-third from the inner end.[39] See Figure 4.1.

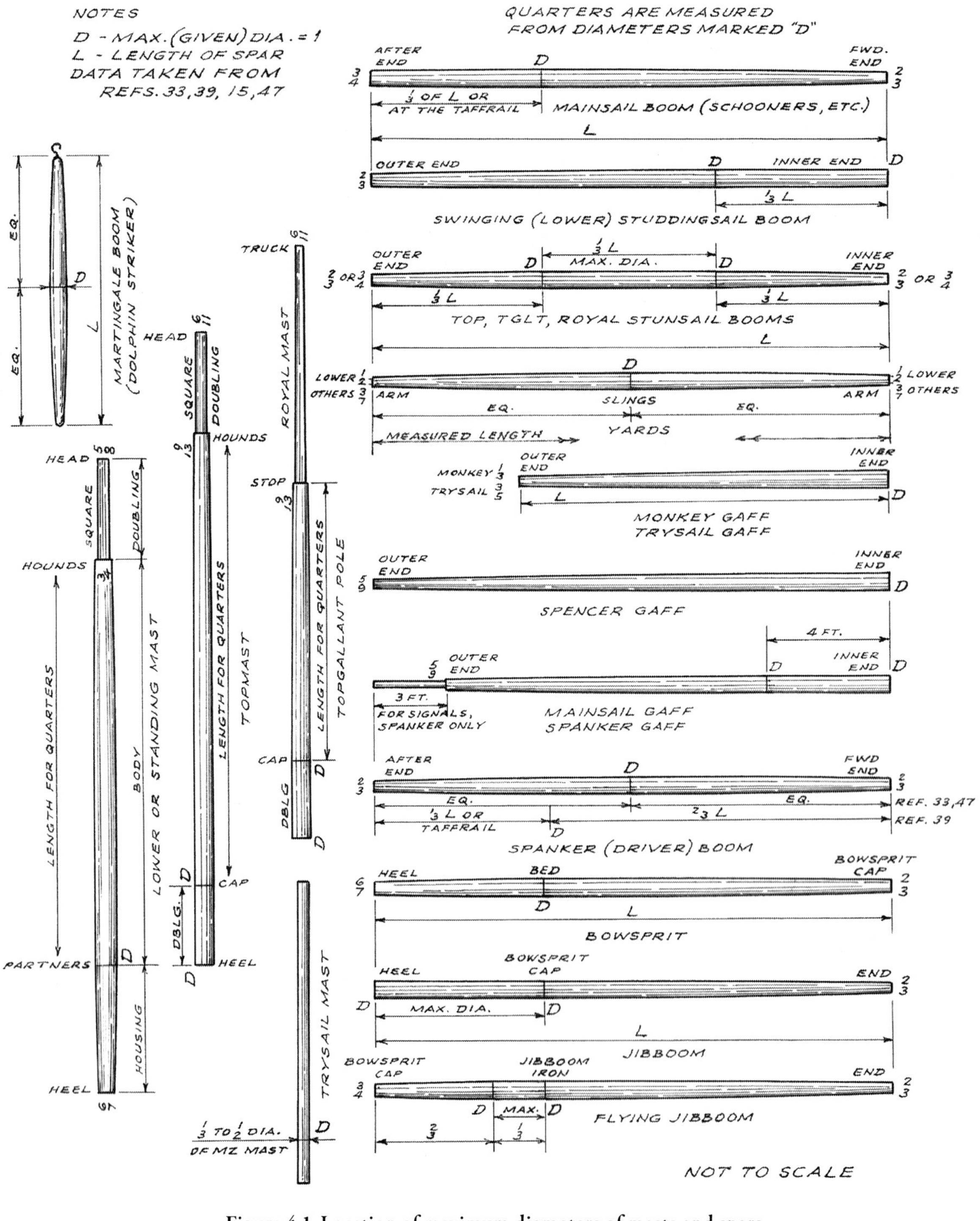

Figure 4.1. Location of maximum diameters of masts and spars.

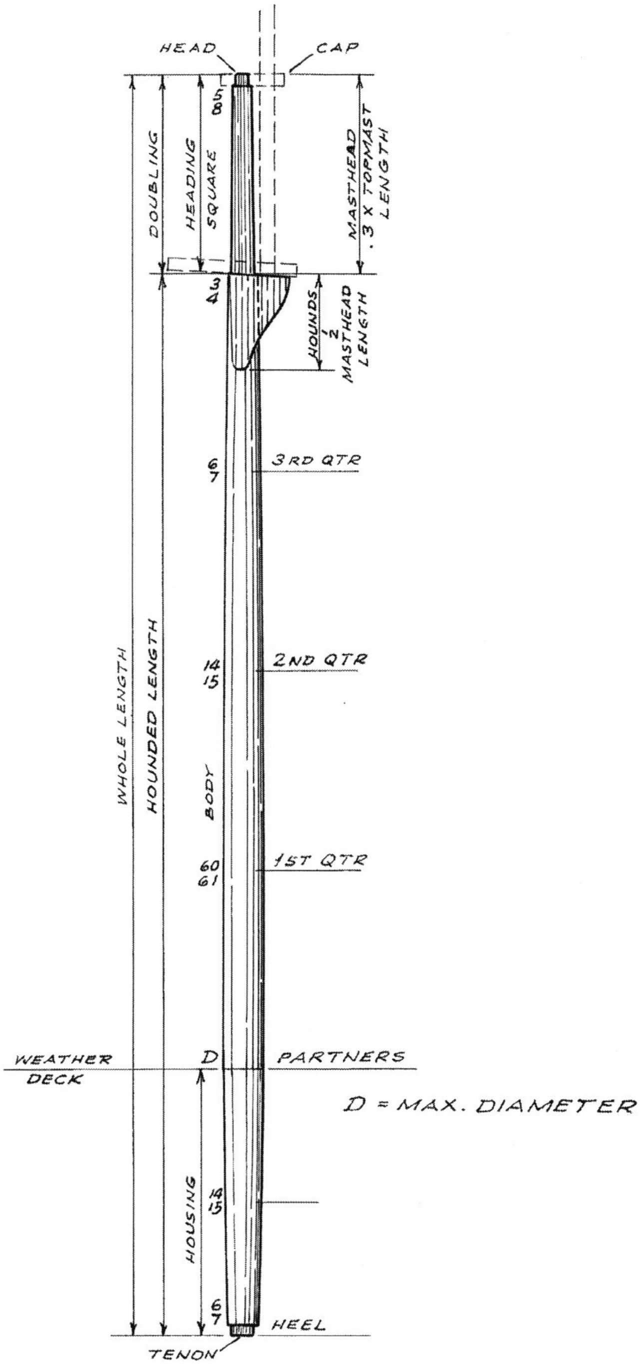

Figure 4.2. Lower mast for single topsails.

The proportions and terminology of a lower, or standing, mast are shown in Figure 4.2.

In Europe the Swedes, Danes, French and English all formulated their own individual set of rules for sparring ships. Tables were developed which gave the proportionate sizes for each spar encountered in a ship.

The English System is recorded in Table 4.1. It is obvious, from reading the entries in this table that proportion in reference to the ship's hull was the constant that governed the sizing of ships' spars.

The English masting table contained no entries for locating the masts along the deck. The Swedish, Danish and French masting tables did, however, include this information which is repeated herewith as details of interest.[15]

Swedish System (distance between p.p., known)
Foremast = 4/31 aft of forward perpendicular.
Mainmast = 2/31 aft of middle of ship.
Mizzenmast = .182 × length of ship forward of aft perpendicular.

Danish System (length of load waterline, known)
Foremast = center 1/8 to 1/9 aft of forward end of LWL.
Mainmast = center 1/14 to 1/15 aft of center of LWL.
Mizzenmast = center 1/6 to 4/25 forward of aft end of LWL.

French System (length of ship on deck, known)
Foremast = .29 × length of ship forward of the middle.
Mainmast = .155 × length of ship aft of middle.
Mizzenmast = .365 × length of-ship aft of middle.

Mast locations that are known for ships included in this book appear in Tables 10.1 and 10.2.

In addition to the above data various other proportionate values were established for spars of British ships of all merchant classes from ships to schooners. These are recorded in Tables 4.2 through 4.10.

Table 4.2 refers to the relationship between the given diameter and the length of the various masts, yards, booms and gaffs.

Table 4.3 lists the proportionate values of the smallest diameter of a spar in reference to the largest, or given, diameter. The largest diameter is given a value of 1. The intermediate diameters represent a fraction of 1.

Table 4.4 lists the fractional values of the intermediate quarters in reference to a given diameter of 1. This includes all categories of spars.

Table 4.5 enumerates the actual diameters of masts at the intermediate quarters and ends, for masts where actual given diameters are known.

Table 4.6 duplicates Table 4.5 for topmasts, topgallants and royal masts. (See note at end of Chapter 4.)

Table 4.7 duplicates Table 4.5 for bowsprits.

Table 4.8 duplicates Table 4.5 for yards.

Table 4.9 duplicates Table 4.5 for driver (spanker), booms, jibbooms, etc.

Table 4.10 duplicates Table 4.5 for main booms and gaffs in fore-and-aft rigged vessels.

According to Fincham[39] the information included in Tables 4.5 through 4.10 was published in about 1807 and this allowed mastmakers and spar makers to read the intermediate quarter diameters directly without making their own calculations using the fractional values.

Table 4.1.
English System of Sparring Ships[15]

L = length between stem and sternpost on deck; B = breadth of ship outside of wales

Mainmast (whole length) = $\frac{L + B}{2}$; diameter = 7/8" to 3 feet of length
Foremast = 8/9 of mainmast
Mizzenmast = 3/4 of mainmast; diameter = 2/3 of mainmast
Main topmast = 3/5 of mainmast; diameter = 1" to 3 feet of length
Fore topmast = 8/9 of main topmast
Mizzen topmast = 5/7 of main topmast; diameter = 7/10 of main topmast
Topgallant mast = 1/2 of topmast; diameter = 1" to 3 feet of length
Royal mast = 3/4 of topgallant mast; diameter = 2/3 of topgallant mast
Bowsprit, whole length = 3/7 of mainmast; outboard = 3/4 its length
Diameter = foremast
Jibboom, outboard = same as bowsprit, outboard; diameter = 1" to 2½ feet
Flying jibboom, outboard = 5/7 of jibboom; diameter = 7/8" to 3 feet of length
Main yard, length = 7/8 of mainmast; diameter = 7/10" to 3 feet of length
Fore yard, length = 7/8 of mainmast
Crossjack = same as fore topsail yard; diameter = 5/8" to 3 feet
Main topsail yard = 5/7 of main yard; diameter = 5/8" to 3 feet
Fore topsail yard = 7/8 of main topsail yard
Mizzen topsail yard = 2/3 of main topsail yard
Topgallant yard = 3/5 of topsail yard (each mast)
Royal yard = 1/2 of topsail yard (each mast)
Mizzen (spanker) boom = same as main topsail yard
Mizzen (spanker) gaff = 5/6 of spanker boom; diameter = 5/8" to 3 feet

Table 4.2.
Proportions of Diameters to the Lengths of Masts and Spars[33]

Mast diameters
Main and fore, ships—1 inch for every 3 feet of length
Mainmast, brigs—1 inch for every 3 feet length
Foremast, brigs—9/10 diameter of mainmast
Mizzenmast, ships—2/3 diameter of mainmast
Main and fore topmasts—1 inch for every 3 feet length
Mizzen topmasts—7/10 diameter of mainmast
Topgallant masts—1 inch for every 3 feet of length
Royal masts—2/3 diameter of their topgallants
Bowsprits—same diameter as mainmast

Yard diameters
Main and fore yards, at slings—7/10 inch to 7/8 inch for ever 3 feet length
Topsail yards at middle—5/8 inch to 7/8 inch for every 3 feet length
Topgallant yards—6/10 inch to 5/8 inch for every 3 feet length
Royal yards—1/2 diameter of their topsail yards
Studdingsail yards—1 inch diameter every 5 feet length

Boom diameters
Jibbooms—7/8 inch for every 3 feet length
Flying jibbooms—7/8 inch for every 3 feet length
Spanker boom—same as fore topsail yard
Gaffs—same as their booms
Studdingsail booms—1 inch every 5 feet length

These tables emphasize the importance of proportion in this phase of shipbuilding.

No exact origin of the fractional values of intermediate quarters of masts and yards, as listed in Table 4.4, can be identified. As shown in the table, the list appears in Steel,[47] originally published in 1795, without additional data. The fractions have a profuse array of denominators and no attempt appears to have been made to modify the fractions so that they had a least common denominator.

Lower masts and yards were not fabricated with straight tapers between their given diameter and the least diameter. The profiles of all have a decided curved surface between the above extremes.

Fincham[39] writes, "Several practical methods are used for determining the intermediate diameters or forms of the masts, etc. between the given or greatest, and the least diameter. Most of the rules in common use give a curve, which approaches nearer to an arc of a circle than to any other known curve, and which appears from experience to comprise the proper degree of substance and strength."

Table 4.4, taken directly from Steel,[47] contains the fractional values of the intermediate quarters of all groups of spars encountered in square rigged and fore-and-aft rigged vessels. From these values it is possible to obtain the nominal diameter of any size spar at the intermediate quarters when a given diameter is known. This is accomplished under the Rule of Three which reads, "When there are three numbers given, if the second be multiplied by the third, and the product divided by the first, the quotient will give the fourth term, which will have the same proportion to the third that the second term has to the first." In our case the fourth term will be the diameter of the spar at the appropriate quarter.

Put another way, if the given diameter of the spar is multiplied by the numerator of the fraction and the product is divided by the denominator of the fraction, the quotient will be the diameter of the spar at the appropriate quarter.

Table 4.3. Proportions of Small Diameters to Large Diameters for Ships Through Schooners[39]

	Masts	*Proportions (Large diameter = 1)*	
Ships, barks, brigs	Standing masts	⅚ at the heel	
		⅘ at the hounds (single tree)	
		¾ at the hounds (cheeked masts)	
		⅔ at the head	
Schooners	Standing masts	⅚ at the heel	
		6/10 or ¾ at the hounds	
		½ or ⅔ at the head	
Ships, barks, brigs	Topmasts	⅘ or 5/7 at hounds	or as large as
		9/13 or 5/9 at head	topgallant masts
Schooners	Topmasts	10/13 or ⅚ at the hounds	
		½ or 3/7 at the head	
Ships, barks, brigs	Topgallant masts	10/13 or ⅘ at the hounds	
		½ or 5/9 at the skysail pole	
Ships, barks, brigs	Bowsprit	⅚ at the heel	
		⅔ at outer end	
Schooners	Bowsprit	1 at the heel	
		⅔ at outer end	
Ships, barks, brigs	Lower yard	½ at yardarm	
Topsail		½ at yardarm	
Topgallant		3/7 or ½ at yardarm	
Royal		3/7 or ½ at yardarm	
Crojack		3/7 or ½ at yardarm	
Studdingsail yard		3/7 or ½ at yardarm	
Schooners	Square sail	½ at yardarm	
Topsail		3/7 at yardarm	
	Booms		
Spanker (Driver)		¾ at outer end; ⅔ at inner end	
Jibboom (separate)		⅔ or ¾ at outer end; 1 at inner end	
Jibboom and flying jibboom in one		⅔ at stops; ½ at outer end	
Studdingsail swinging boom (lower)		⅔ or ¾ at outer end and inner end	
	Gaffs		
Ships, barks, brigs		½ or 5/9 at outer end	
Schooners		⅗ or 5/9 at outer end	
Trisail gaffs		⅗ or 5/9 at outer end	

Table 4.4. Fractional Proportions of the Intermediate Quarters to the Given Diameter of Masts and Spars[33,47]

Spar	*Quarters*			*Head (Sq.), lower and topmasts only*		
	1st	*2nd*	*3rd*	*Lower part*	*Upper part*	*Heel*
Standing (lower) masts	60/61	14/15	6/7	¾	⅝	6/7
Topmasts, topgallant masts, royal masts	60/61	14/15	6/7	9/13	6/11	
Yards	30/31	⅞	7/10	Arms 3/7		
Bowsprit	60/61	11/12	⅘	Outer end 5/9		6/7
Jibboom, spanker boom, flying jibboom	40/41	11/12	⅚	Ends ⅔		
Main booms	40/41	12/13	⅞	Fore end ⅔	Middle 11/12	Aft end ¾
Gaffs	40/41	11/12	⅘	End 5/9		
Heeling—Standing masts	⅔ athwartship		½ fore and aft			
Heeling—Bowsprit	7/12 athwartship		⅔ up and down			

Table 4.5.
Diameters of Lower Masts at Their Quarters, Heads and Heels[33,47]

Dia. in inches at partners	*Quarters* 1st	*Quarters* 2nd	*Quarters* 3rd	*Head (Sq.) Lower part* Thwart	*Head (Sq.) Lower part* F&A	*Head (Sq.) Upper part*	*Heels*
	60/61	14/15	6/7	6/7	¾	⅝	6/7
44	43¼	41	37¾	37¾	33	27½	37¾
43½	42¾	40⅝	37¼	37¼	32⅝	27 3/16	37¼
43	42¼	40⅛	36⅞	36⅞	32¼	26⅞	36⅞
42½	41¾	39⅝	36½	36½	31⅞	26 9/16	36½
42	41¼	39¼	36	36	31½	26¼	36
41½	40¾	38¾	35⅝	35⅝	31⅛	25 15/16	35⅝
41	40⅜	38¼	35⅛	35⅛	30¾	25⅝	35⅛
40½	39⅞	37⅞	34¾	34¾	30⅜	25 5/16	34¾
40	39⅜	37⅜	34⅜	34⅜	30	25	34⅜
39½	38⅞	36⅞	34	34	29⅝	24¾	34
39	38⅜	36⅜	33½	33½	29¼	24⅜	33½
38½	37⅝	35⅞	33	33	28⅞	24	33
38	37⅜	35⅜	32½	32½	28½	23¾	32¼
37½	36⅞	35	32⅛	32⅛	28⅛	23⅜	32⅛
37	36⅜	34½	31¾	31¾	27¾	23⅛	31¾
36½	35⅞	34	31⅜	31⅜	27⅜	22⅞	31⅜
36	35⅜	33⅝	30⅞	30⅞	27	22½	30⅞
35½	34⅞	33⅛	30½	30½	26⅝	22¼	30½
35	34⅜	32⅝	30	30	26¼	21⅞	30
34½	33⅞	32⅛	29½	29½	25⅞	21½	29½
34	33⅜	31⅝	29⅛	29⅛	25½	21¼	29⅛
33½	32⅞	31⅛	28¾	28¾	25⅛	21	28¾
33	32½	30¾	28¼	28¼	24¾	20⅝	28¼
32½	31⅞	30¼	27⅞	27⅞	24⅜	20¼	27⅞
				Hounds			
32	31½	29¾	27½	24		20	27½
31½	31	29⅜	27	23⅝		19¾	27
31	30½	28⅞	26⅝	23¼		19½	26⅝
30½	30	28½	26	22⅞		19⅛	26
30	29½	28	25¾	22½		18⅞	25¾
29½	29	27½	25¼	22⅛		18½	25¼
29	28½	27	24⅞	21¾		18⅛	24⅞
28½	28	26½	24⅜	21⅜		17¾	24⅜
28	27½	26⅛	24	21		17½	24
27½	27	25⅝	23½	20⅝		17⅛	23½
27	26½	25¼	23⅛	20¼		16⅞	23⅛
26½	26	24¾	22¾	19⅞		16⅝	22¾
26	25½	24¼	22¼	19½		16¼	22¼
25½	25	23¾	21⅞	19⅛		15⅞	21⅞
25	24½	23¼	21⅜	18¾		15⅝	21⅜
24½	24	22¾	21	18⅜		15¼	21
24	23½	22⅜	20½	18		15	20½
23½	23⅛	21⅞	20⅛	17⅝		14¾	20⅛
23	22⅝	21½	19¾	17¼		14½	19¾
22½	22⅛	21	19¼	16⅞		14⅛	19¼
	60/61	14/15	6/7	6/7	¾	⅝	6/7
				Hounds			
22	21⅝	20½	18⅞	16½		13¾	18⅞
21½	21⅛	20	18⅜	16⅛		13½	18⅜
21	20⅝	19⅝	18	15¾		13⅛	18
20½	20⅛	19⅛	17⅝	15⅜		12¾	17⅝
20	19¾	18¾	17⅛	15		12½	17⅛
19½	19⅛	18⅛	16¾	14⅝		12⅛	16¾
19	18¾	17¾	16 ¼	14¼		11⅞	16¼
18½	18⅛	17¼	15 ¾	13⅞		11⅝	15¾
18	17¾	16¾	15 ½	13½		11½	15½
17½	17⅛	16¼	15	13⅛		11	15
17	16¾	15⅞	14 ½	12¾		10⅝	14½
16½	16⅛	15⅜	14 ⅛	12⅜		10⅜	14⅛
16	15¾	14⅞	13 ¾	12		10	13¾
15½	15¼	14½	13 ¼	11⅝		9¾	13¼
15	14¾	14	12 ⅞	11¼		9⅜	12⅞
14½	14¼	13½	12 ½	10⅞		9⅛	12½
14	13¾	13	12	10½		8¾	12
13½	13¼	12¼	11 ½	10⅛		8½	11½
13	12¾	12⅛	11 ⅛	9¾		8⅛	11⅛
12½	12¼	11⅝	10 ⅝	9⅜		7⅞	10⅝
12	11¾	11⅛	10 ¼	9		7½	10¼

Three methods of producing the desired curve were known to be in use in the 1850s. They are described as follows, and are illustrated in Figure 4.3. All are taken from Fincham[39] under the heading "The Diameters or Forms of Masts, Yards, etc. between the Given or Large, and the Small Diameters."

Royal dockyards method: Divide the distance between the location of the given diameter, and that of the smallest diameter, into four equal parts, as from the partners or caps, to the hounds or stops, which in lower or standing masts are called the upper quarters, and from the partners to the heel, which are called the lower quarters. These quarters are identified as 1, 2 and 3. The difference between the given diameter and the proportion for the heel, hounds or yardarms, etc., is divided into four parts; for the third quarter it is half the difference, the second one-fifth, and the first the difference less one-fifteenth of the diameter. This method is illustrated in Figure 4.3, Detail 1.

Mastmakers method: Mark from one end of a batten, called a graduating batten, a distance equal to one-half the given diameter, and another equal to half this portion for the heel, hounds, etc. (One-half is taken because all diameters are set off from a middle line, which represents the axis, and thus the form of both sides of the mast is made the same.) Then the difference between these two distances is divided into two equal parts, which gives one-half the distance for the third quarter, two-fifths of this one-half from the given diameter for the second quarter, and one-third of the two-fifths of the one-half for the first quarter. This method is illustrated in Figure 4.3, Detail 2.

The French method: This method should be read in conjunction with Figure 4.3, Detail 3. Draw a line AB, equal to one-half the greatest diameter; from A and B as centers, with the radius AB, scribe two arcs intersecting at P. Bisect line AB at C, and on each side of C take CD and CE, each equal to one-fourth of the smallest diameter; from D and E raise perpendiculars DF and EG, intersecting the arcs at F and G; connect F and G. Then FG is equal to DE, which is equal to half the smallest diameter.

Table 4.6. Diameters of Topmasts, Topgallants and Royals, at Their Quarters and Heads[33,47]

Dia. in the cap (inches)	Quarters 1st	Quarters 2nd	Quarters 3rd	Head (Sq., topmast only) Lower part	Head (Sq., topmast only) Upper part
	60/61	14/15	6/7	9/13	6/11
24	23⅝	22⅜	20 9/16	16⅝	13⅛
23½	23⅛	21 15/16	20⅛	16¼	12¾
23	22⅝	21 7/16	19¾	15⅞	12½
22½	22⅛	21	19¼	15 9/16	12¼
22	21⅝	20½	18⅞	15¼	12
21½	21⅛	20	18⅜	14⅞	11¾
21	20⅝	19⅝	18	14½	11½
20½	20⅛	19⅛	17⅝	14⅛	11⅛
20	19¾	18¾	17⅛	13⅞	10⅞
19½	19⅛	18⅛	16¾	13½	10⅝
19	18¾	17¾	16¼	13⅛	10⅜
18½	18⅛	17¼	15¾	12⅞	10⅛
18	17¾	16¾	15½	12½	9⅞
17½	17⅛	16¼	15	12⅛	9½
17	16¾	14⅞	14½	11¾	9¼
16½	16⅛	15⅜	14⅛	11⅜	9
16	15¾	14⅞	13¾	11 1/16	8¾
15½	15¼	14½	13¼	10¾	8½
15	14¾	14	12⅞	10⅜	8¼
14½	14¼	13½	12½	10	7⅞
14	13¾	13	12	9¾	7⅝
13½	13¼	12½	11½	9⅜	7⅜
13	12⅜	12⅛	11⅛	9	7⅛
12½	12¼	11⅝	10⅝	8⅝	6⅞
12	11¾	11⅛	10¼	8 5/16	6⅝
11½	11¼	10⅝	9⅞	7 15/16	6¼
11	10¾	10¼	9½	7⅝	6
10½	10¼	9¾	9	7¼	5¾
10	9¾	9¼	8⅝	6 15/16	5½
9½	9¼	8⅞	8⅛	6 9/16	5¼
9	8⅞	8⅜	7¾	6¼	5
8½	8⅜	7⅞	7¼	5 15/16	4⅝
8	7⅞	7 7/16	6⅞	5½	4⅜
7½	7⅜	7	6⅜	5⅛	4⅛
7	6⅞	6½	6	4⅞	3⅞
6½	6⅜	6 1/16	5⅝	4½	3½
6	5⅞	5⅝	5⅛	4⅛	3¼
5½	5⅜	5⅛	4¾	3 13/16	3
5	4⅞	4⅝	4¼	3 7/16	2¾
4½	4⅜	4⅛	3⅞	3⅛	2½
4	3⅞	3¾	3⅜	2¾	2¼

This being done divide the length of mast which exists between the greatest and least diameters into four equal parts; then divide the space between FG and DE into a like number of equal parts; extend these parts with lines parallel with base AB until they meet arcs AP and BP. Designate these new lines as HL, IM and KN. These new lines will be half diameters of the mast at the intermediate quarters.

If these three methods are laid out to the same scale, and developed as directed, the resultant curves will match exactly if one is overlaid upon another.

The proportions of spars, as they appeared in the 1850s, were the result of years of study and observation. Various attempts at developing formulae which would predict results of certain conditions and stresses proved unsuccessful. Ultimately a record of previous sizes, materials and designs led to the formation of tables based on past experience.

Three basic criteria were established as being necessary to fashion an acceptable mast, bowsprit, yard or boom. These criteria were strength, weight and appearance.

The most obvious requirement was that of strength. Masts were burdened with great dead loads of the masting above, along with stresses induced by the actions of a ship at sea under extreme weather conditions. Yards were subject to great and variable wind loads when all sail was set. Bowsprits, in addition to the loads induced by the headstays which they supported, were subject to severe pounding by headseas in heavy weather.

By their physical characteristics pine and fir trees were the most suitable spar timbers. The trees grew in spectacular, straight growth in densely forested areas. This growth resulted in tree trunks with tremendously long, straight grain which was essential to the construction of masts and spars.

These evergreens were composed of wood where strength was consistent throughout the individual species, and this strength had been studied and catalogued so that all the qualities were fully known. With the knowledge at hand the pines and firs became the stuff of masts and spars. The inherent strength of the wood governed the size of timbers to be used in all cases. As a result countless tables were formulated for various applications. This information allowed ships to be built in widely separated shipyards but all similar because of the standards.

The second consideration, that of weight, was of utmost importance but had little to do with the actual weight of any given log. It was the potential location of this log that was important.

By the nature of sailing ships the function of a mast is to raise a yard to some distance above the deck. In turn, another mast, smaller than the lower mast, raised another, even smaller, yard, and so on. It was essential that each mast or yard be sufficiently strong but smaller in size so as to not compromise the stability of the ship. The center of gravity of these mast and yard assemblies was high above the deck and was a great load to be considered.

The magnitude of such a load may be appreciated when noting the weights encountered in the clipper packet *Donald McKay*,[1] a very large vessel but one which allows comparison with smaller ships. As recorded in Ref. 1, the following weights are noted. The lower masts, fore, main and mizzen, 30, 33, 18 tons respectively; topmasts, fore, main and mizzen,

Table 4.7.
Diameters of Bowsprits at Their Quarters and Ends[33,39]

Diameter at the Bed (inches)	*Quarters* 1st	*Quarters* 2nd	*Quarters* 3rd	*Ends* Outer End	*Ends* Heel
	60/61	11/12	4/5	2/3	6/7
44	43¼	40⅜	35¼	29 5/16	37¾
43½	42¾	39⅞	34¾	29	37¼
43	42¼	39⅜	34⅜	28⅝	36⅞
42½	41¾	39	34	28 5/16	36½
42	41¼	38½	33⅝	28	36
41½	40¾	38	33¼	27⅝	35⅝
41	40⅜	37⅝	32¾	27 5/16	35⅛
40½	39⅞	37⅛	32⅜	27	34¾
40	39⅜	36⅝	32	26⅝	34⅜
39½	38⅞	36⅛	31⅝	26 5/16	34
39	38⅜	35¾	31¼	26	33½
38½	37⅝	35¼	30⅞	25⅝	33
38	37⅜	34⅞	30⅜	25¼	32½
37½	36⅞	34⅜	30	25	32⅛
37	36⅜	34	29½	24⅝	31¾
36½	35⅞	33½	29⅛	24¼	31⅜
36	35⅜	33	28¾	24	30⅞
35½	34⅞	32⅝	28⅜	23⅝	30⅜
35	34⅜	32⅛	28	23⅜	30
34½	33⅞	31⅝	27⅝	23	29½
34	33⅜	31⅛	27¼	22⅝	29⅛
33½	32⅞	30⅝	26¾	22¼	28¾
33	32½	30¼	26⅜	22	28¼
32½	31⅞	29¾	26	21⅝	27⅞
32	31½	29⅜	25⅝	21⅜	27½
31½	31	28⅞	25¼	21	27
31	30½	28½	24¾	20⅝	26⅝
30½	30	28	24⅜	20⅜	26
30	29½	27½	24	20	25¾
29½	29	27	23⅝	19⅝	25¼
29	28½	26⅝	23¼	19⅜	24⅞
28½	28	26	22¾	19	24⅜
28	27½	25⅝	22⅜	18⅝	24
27½	27	25¼	22	18⅜	23½
27	26½	24¾	21⅝	18	23⅛
26½	26	24¼	21¼	17⅝	22¾
26	25½	23⅞	20¾	17 9/16	22¼
25½	25	23⅜	20⅜	17	21⅞
25	24½	22⅞	20	16⅝	21⅜
24½	24	22½	19⅝	16 9/16	21
24	23½	22	19¼	16	20½
23½	23⅛	21½	18¾	15⅝	20⅛
23	22⅝	21⅛	18⅜	15 9/16	19¾
22½	22⅛	20⅝	18	15	19¼
22	21⅝	20⅛	17½	14⅝	18⅞
21½	21⅛	19¾	17⅛	14 9/16	18⅜
21	20⅝	19¼	16¾	14	18
20½	20⅛	18⅞	16⅜	13 11/16	17⅝
20	19¾	18¼	16	13 5/16	17⅛
19½	19⅛	17⅞	15⅝	13	16¾
19	18¾	17½	15⅛	12 11/16	16¼
18½	18⅛	17	14¾	12¾	15¾
18	17¾	16¾	14⅜	12	15½
17½	17⅛	16	14	11 11/16	15
17	16¾	15½	13⅝	11¼	14½
16½	16⅛	15⅛	13⅛	11	14⅛
16	15¾	14⅝	12¾	10⅝	13¾
15½	15¼	14⅛	12⅜	10¼	13¼
15	14¾	13⅞	12	10	12⅞
14½	14¼	13¼	11⅝	9¾	12½
14	13¾	12⅞	11⅛	9⅜	12
13½	13¼	12⅜	10¾	9	11½
13	12¾	11⅞	10⅜	8¾	11⅛
12½	12¼	11½	10	8¼	10⅝

5, 5, 3 tons, respectively; bowsprit, 7 tons. The lower yards 12, 14, 6 tons; lower topsail yards, 8, 8, 4 tons; upper topsail yards, 4, 5, 3 tons. A total of 167 tons of pitch pine. In addition there were 30, 31 and 28 iron hoops on her lower masts, weighing 8830 lbs.; hoops on the fore and main yards, 3890 lbs.; and lower yard trusses, 2250 lbs.

Over a period of time it was found that lower masts, and yards were each subject to their own particular defects. In the case of masts they tended to wring or twist at the head above the hounds. Yards were commonly condemned because of shans or strains due to the taper of the yard crossing the fiber or grain of the wood and exiting from the surface of the yard near the yardarm. Bowsprits had their own prevalent defect—that of being sprung between the gammoning and the cap.[39]

Appearance and proportion were the final ingredients which were worked into the production of a ship's spars.

There is an old saying in structural design work, "If it does not look right, it probably is not right. Check it." It was very appropriate in the building of wooden merchant vessels, especially after the finely formed and elegant clippers appeared.

In the masting of ships strength and weight were the two primary elements to be considered but, as these were being constantly controlled, the ship designers and builders, and even the cost-conscious owners, were very attentive to the general appearance of the vessels they produced. This is very apparent in a visual inspection of a three masted, square rigged vessel.

The lower masts, protruding through the partners at the weather deck, were solid, sturdy structures which lent them-

Table 4.8.
Diameters of Yards at Their Quarters[33,47]

Diameter at the Slings (inches)	*Quarters* *1st*	*2nd*	*3rd*	*Arm*
	30/31	⅞	7/10	3/7
28	27	24½	19⅗	12
27½	26⅝	24	19¼	11⅞
27	26⅛	23⅝	19	11 9/16
26½	25⅝	23 3/16	18½	11⅜
26	25 3/16	22¾	18¼	11 3/16
25½	24 11/16	22 3/16	17⅞	11
25	24⅛	21⅞	17½	10¾
24½	23⅝	21⅜	17⅛	10½
24	23¼	21	16¾	10¼
23½	22¾	20 9/16	16½	10 1/16
23	22¼	20⅛	16 1/16	9⅞
22½	21¾	19¾	15¾	9⅝
22	21¼	19¼	15⅜	9 7/16
21½	20¾	18⅞	15	9¼
21	20¼	18⅜	14¾	9
20½	19⅞	17 15/16	14⅜	8¾
20	19⅜	17½	14	8 9/16
19½	18⅞	17	13⅝	8¾
19	18½	16½	13¼	8⅛
18½	17 15/16	16⅛	13	8
18	17½	15¾	12⅝	7¾
17½	17	15¼	12¼	7½
17	16½	14⅞	11⅞	7¼
16½	16	14⅜	11⅝	7 1/16
16	15½	14	11 3/16	6⅞
15½	15	13 9/16	10⅞	6⅝
15	14½	13⅛	10½	6½
14½	14	12¾	10⅛	6¼
14	13½	12¼	9¾	6
13½	13	11⅞	9½	5¾
13	12½	11¼	9 1/16	5 9/16
12½	12 1/16	10⅞	8¾	5⅜
12	11⅝	10½	8⅜	5⅛
11½	11 1/16	10	8	4 15/16
11	10⅝	9⅝	7¾	4¾
10½	10 1/16	9⅛	7⅜	4½
10	9⅝	8¾	7	4¼
9½	9⅛	8¼	6⅝	4 1/16
9	8¾	7⅞	6¼	3⅞
8½	8¼	7⅜	5 15/16	3⅝
8	7¾	7	5⅛	3⅜
7½	7¼	6½	5¼	3¼
7	6¾	6⅛	4 15/16	3
6½	6¼	5¾	4½	2¾
6	5⅞	5¼	4 3/16	2 9/16
5½	5 5/16	4⅞	3⅞	2⅜
5	4⅞	4⅜	3½	2 ⅛
4½	4⅜	3 15/16	3⅛	1 15/16
4	3⅞	3½	2 13/16	1¾
3½	3⅜	3 1/16	2 7/16	1½

Table 4.9.
Diameters of Spanker Booms and Jibbooms at Their Quarters[33]

Given Diameter (inches)	*Quarters* *1st*	*2nd*	*3rd*	*Arm*
	10/41	11/12	⅚	⅔
16	15⅝	14⅝	13⅜	10⅝
15½	15 1/16	14⅛	13	10¼
15	14⅝	13¾	12½	10
14½	14 1/16	13¼	12	9¾
14	13¾	12⅞	11⅝	9⅜
13½	13⅛	12⅜	11¼	9
13	12¾	11 15/16	10⅞	8¾
12½	12⅛	11½	10 7/16	8¼
12	11¾	11	10	8
11½	11⅛	10½	9⅝	7⅝
11	10¾	10 1/16	9⅛	7⅜
10½	10¼	9⅝	8¾	7
10	9¾	9⅛	8⅜	6⅝
9½	9¼	8¾	7 15/16	6¼
9	8¾	8¼	7½	6
8½	8¼	7¾	7 1/16	5⅝
8	7⅞	7⅜	6⅝	5⅜
7½	7⅜	6⅞	6¼	5
7	6⅞	6⅜	5⅞	4¾
6½	6⅜	6	5 7/16	3⅜
6	5⅞	5½	5	4
5½	5⅜	5 1/16	4⅝	3 11/16
5	4⅞	4⅝	4⅛	3⅜
4½	4⅜	4⅛	3¾	3
4	3⅞	3⅝	3⅜	2⅝
3½	3 7/16	3¼	2 15/16	2⅜

selves harmoniously to the great bulk of the hull which supported them.

The masts that rose successively above the standing masts were markedly more slender but they decreased in diameter and height in an orderly and graceful progression. At their termination they were capped with a small truck.

As the masts rose above the deck the yards that crossed them were fashioned in a corresponding progression. The lowest yard was a long, sturdy spar which was proportioned in the manner of its supporting mast.

The higher yards were successively shorter and more slender, in keeping with the proportions of their supporting masts.

In many ships the lengths of the yards as they were crossed, starting with the lowest and ending with the highest, generally the royal yard, were proportioned so that, when all sail was set, the leeches of the successive sails stretched in a straight line from the topmost earing (of the royal sail) to the lower-most clew (that of the topsail). The symmetry and visual appeal of such an arrangement was not accidental.

Table 4.10. Diameters of Main Booms and Gaffs at Their Quarters[33]

Main Booms

Outer End	Middle	Given Diameter (Inches)	Quarters 1st	Quarters 2nd	Quarters 3rd	Inner End
¾	11⁄12		40⁄41	12⁄13	⅞	⅔
12	14⅝	16	15⅝	14¾	14	10⅝
11⅝	14⅛	15½	15 1⁄16	14¼	13½	10¼
11¼	13¾	15	14⅝	13⅞	13⅛	10
10⅞	13¼	14½	14 1⁄16	13⅜	12⅝	9¾
10½	12⅞	14	13¾	13	12¼	9⅜
10⅛	12⅜	13½	13⅛	12½	11¾	9
9¾	11 15⁄16	13	12¾	12	11⅜	8¾
9⅜	11½	12½	12⅛	11 9⁄16	10⅞	8¼
9	11	12	11¾	11 1⁄16	10½	8
8⅝	10½	11½	11⅛	10 9⁄16	10	7⅝
8¼	10 1⁄16	11	10¾	10⅛	9⅝	7⅜
7⅞	9⅝	10½	10¼	9 11⁄16	9⅛	7
7½	9⅛	10	9¾	9¼	8¾	6⅝
7⅛	8¾	9½	9¼	8¾	8¼	6¼
6¾	8¼	9	8¾	8 5⁄16	7⅞	6
6⅜	7¾	8½	8¼	7 13⁄16	7⅜	5⅝
6	7⅜	8	7⅞	7⅜	7	5⅜

Gaffs

Given Diameter (Inches)	Quarters 1st	Quarters 2nd	Quarters 3rd	End
	40⁄41	11⁄12	⅘	5⁄9
12	11¾	11	9⅝	6⅝
11½	11⅛	10½	9⅛	6⅜
11	10¾	10 1⁄16	8¾	6⅛
10½	10¼	9⅝	8⅜	5⅞
10	9¾	9⅛	8	5 9⁄16
9½	9¼	8¾	7 9⁄16	5¼
9	8¾	8¼	7¼	5
8½	8¼	7¾	6¾	4¾
8	7⅞	7⅜	6⅜	4 7⁄16
7½	7⅞	6⅞	6	4⅛
7	6⅞	6⅜	5 9⁄16	3⅞
6½	6⅜	6	5 3⁄16	3⅝
6	5⅞	5½	4¾	3 5⁄16

Sparse data in masting tables (1850s) indicate that gaff diameters are about three-quarters of given boom diameters.

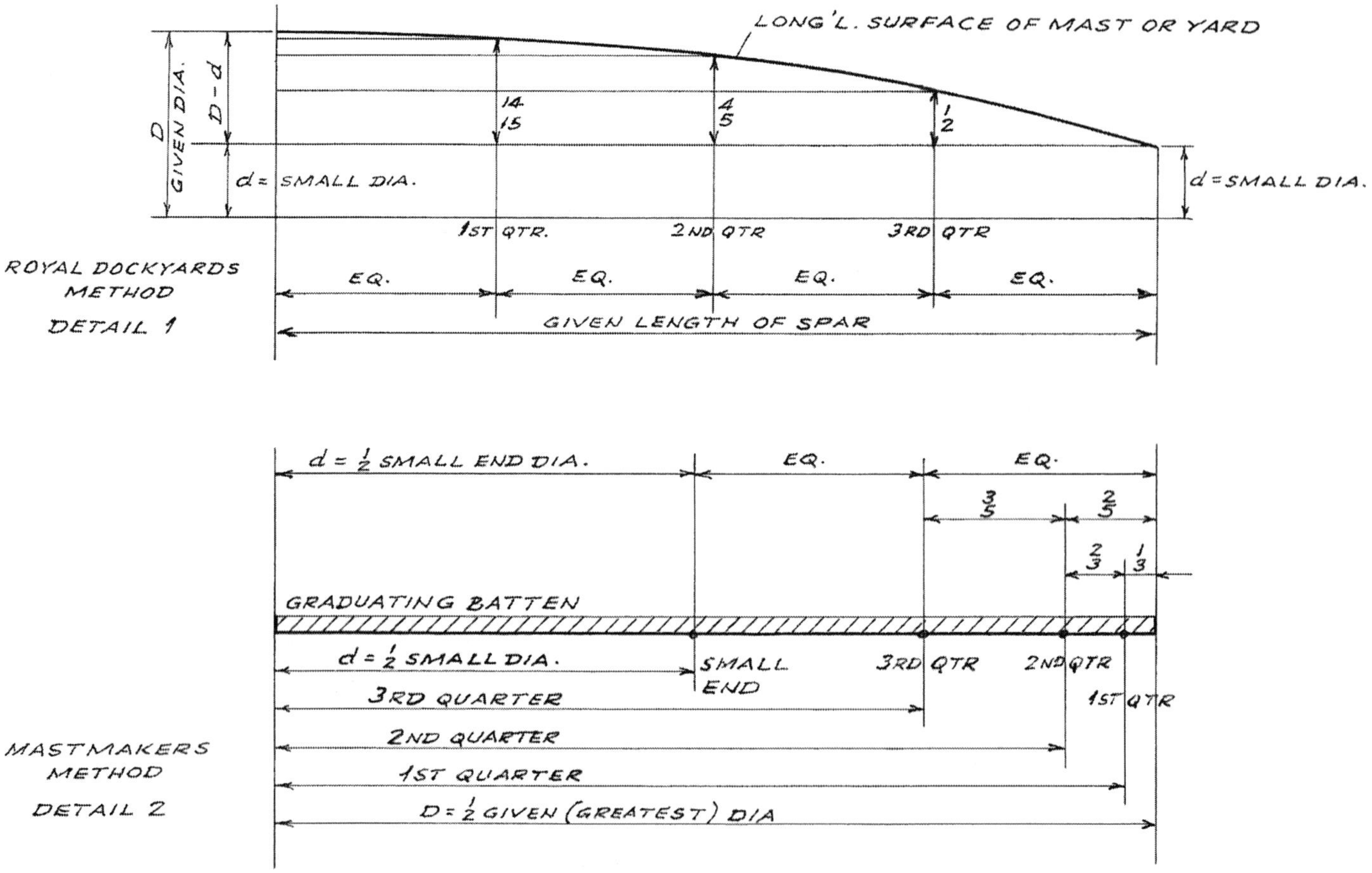

Above and opposite: Figure 4.3. Developing the shape of lower masts and yards.

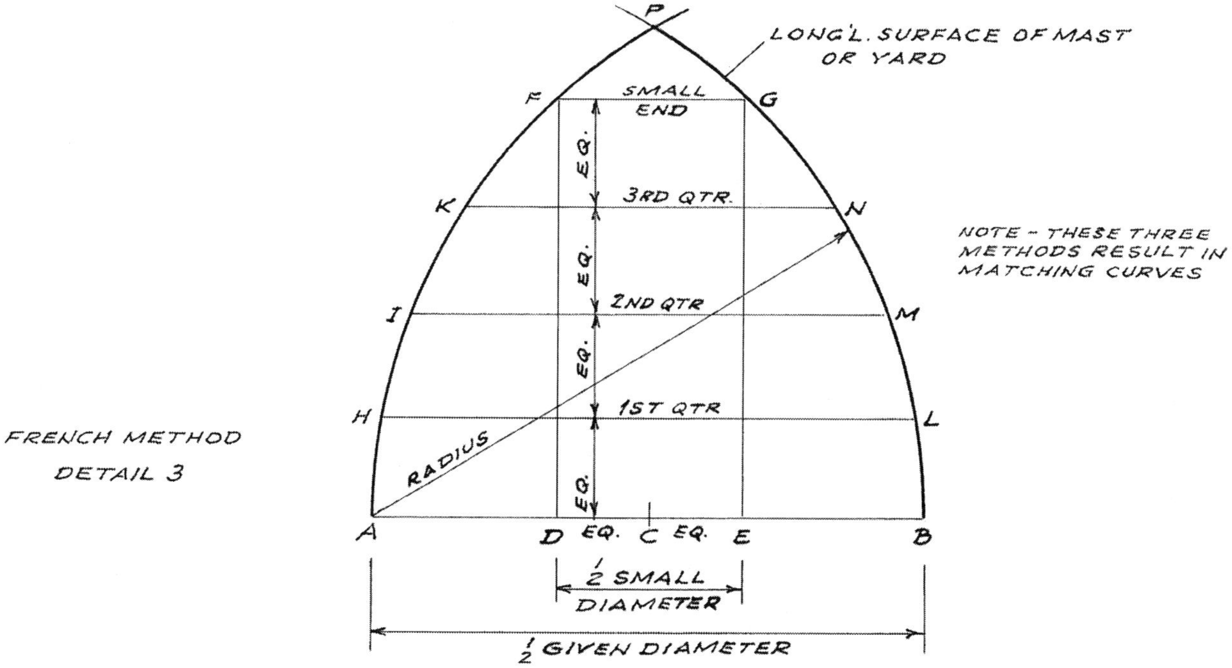

This graceful and harmonious assembly was accomplished while, at all times, each individual mast and yard was sized to be of suitable strength to function as required.

The height (tauntness) of masts was based on the maximum beam of a vessel and was governed by considerations of stability, i.e., the amount of permissible roll of the vessel.

The length (squareness) of yards was based on the length of the vessel on the weather deck and was governed by the consideration of one mast's sails not being so square as to interfere with the wind exerted on the sails of the next mast.

Tables were compiled recording the proportions of masts and yards using data that had been collected through years of experiment and observation.

Note, re: Table 5.6. This table dates from 1794[47] when topgallant and royal masts were headed. In the 1850s these masts, along with the skysail mast, were combined into a single pole mast, not headed.

5

A Single Tree Mast and Bowsprit

In earliest times a mast consisted of a short, unadorned pole jammed against a boat's keel and supported at its top by ropes sufficient to maintain its vertical position.

As time went by boats became larger, masts became longer and the rope supports more numerous. As with all creations an increase in size brought with it an increase in other considerations, such as weight and the effect of outside forces.

Ultimately small boats developed into large ships which required increasingly larger scantlings to preserve their form and increasingly large masts and yards to capture sufficient wind to propel them.

For many centuries lower masts were made from a single tree trunk, but there ultimately came a time when the ships became so large as to outstrip the availability of suitable tree trunks from which to construct these masts.

In the 15th century shipbuilders resorted to fabricating large masts from a collection of small timbers bundled together to form a fasces (from the Latin fascis, meaning a bundle.) This process solved the problem of an inadequate timber supply but was costly and time-consuming albeit very effective. The making of such masts required expert craftsmanship to fashion the individual pieces of timber into a strong and dependable mast. It may have been at this period that mast-making became its own particular craft.

The individual timbers were designed to fit snugly together and not slide out of their allotted location. The staves were bound together into a single strong unit with turns of rope at intervals along the mast. These bindings were called wooldings.

The result was a structure that was superior to the single tree mast in two ways. First, the assembly of staves in the made-mast was stronger than the single tree trunk—the whole was greater than the sum of its parts.

The second trait of superiority lay in the fact that, in assembling the pieces that would compose the final mast, any piece that exhibited a form of decay, splintering or undesirable grain could be detected and discarded.

With the single tree mast the butt and top of an individual log could be visually examined and, if necessary, discarded, but the interior might be rife with hidden defects which an expert could not anticipate.

By the 17th century the making of made-masts had developed into a complicated, sophisticated trade which was carried on only by well-trained craftsmen.

Developing from a simple, small pole of ancient times to the end of the days of sail, the made-mast became the most incredibly intricate piece of wooden construction that sailed in a wooden hull.

The made-mast opened the way for huge hulls to be built. The masts became more and more taunt (the sailors' word for the height of a mast) and the yards became more square (the sailors' word for the length of a yard.)

The only limits to the magnitude of ship construction were now determined by the fiber stresses of the materials that could be used in construction of the whole.

To Make a Single Tree Mast

The spar maker—the term is intended to apply to any individual who worked in a facility that fabricated masts—had first to select a stick which would satisfy the physical requirements of the mast that was to be made.

Green timber, that is, timber that had been freshly harvested, was never used. All masts were made from trees that had been cut and trimmed, and then delivered to a holding basin or, sometimes, to the spar maker, to be properly seasoned, a process that could take many months. See Chapter 3 for details dealing with the preservation of timber.

The various facilities—shipyards, holding basins, spar makers' yards, etc.—kept a detailed inventory of individual sticks that were available for use. Every stick was identified by type of wood and the size of the stick (butt diameter, top diameter and length). The stick diameter is the measurement across the flats of a square of the given size.

All masts were first made square before their parts were rounded.[33,39]

When a mast was to be produced a proper stick was removed from a holding basin or from the sparmaker's ready inventory. The simplest form of requisitioning a required stick occurred when a mast of limited diameter, about 25 inches, was ordered. Tables had been compiled which spelled out the tree diameter required for a given diameter mast and this information was recorded for every stick.

Hewing a finished mast from stock sizes was most economical since much excess material had been removed when the tree was felled.

The first step after a stick of suitable dimension had been chosen was to cut off a cross section of the butt several inches thick. The same was done to the top of the stick. The exposed ends were examined for any signs of decay which, if present, would render the stick unacceptable.

The qualities most sought were a fine, close grain, and layers of ligneous (woody) material that were compact, and concentric annual rings. The compact layers were stronger and had a greater degree of elasticity. The quantity of resin they contained was also a very significant qualification.

The length of the chosen stick included the tenon at the heel of the mast, to seat in the mast step, and the tenon in the masthead to fit into the cap.

Wooden blocks called thawts were placed at strategic locations along the stick at a height to allow it to rest horizontally at about waist height above the ground.

The stick was first trimmed four-square by sawing or hewing with axes. Sawing was the preferred method for reasons of economy. It allowed the cut-off material to be used for fishes and fillings in smaller masts. See Fig. 5.1, lst trim.

After being squared the upper surface of the stick was marked by using a ram line[47] which was a line stretched along the middle for its entire length. Usually this line was secured at one end then stretched taut at the other end. Its trace was then marked on the surface of the squared stick.

After the middle line had been established the locations of the heel, partners, first, second and third quarters, and hounds and head were marked on the surface and lines perpendicular to the middle line were scribed across from one side to the other. Using these lines as guides the half breadths of the mast at the above locations were transferred from the graduating batten to each side of the middle line. After this a line faired through these transverse points, from the heel to the head, indicated the outline of the mast.

The graduating batten is illustrated in Fig. 4.3, detail 2. This operation is illustrated in Fig. 5.1, 2nd trim.

At this time, using a plumb line, a vertical middle line was traced on each end of the stick. Then, from the heel, partners, quarters, and hounds and head, half breadths were cut into its sides until, using the plumb line, the sides had been scored for their full depth. In this manner the mast profile traced on the upper surface could be duplicated on the lower surface. Both sides of the stick were hewed down to these limits, using the scoring on the top surface and the hewed vertical cuts on the sides to produce the profile of the mast. These sides were the forward and after sides of the finished mast. See Fig. 5.1, 3rd trim.

The stick was then canted, or turned, with a trimmed side uppermost. New middle lines were scored on the trimmed upper surface and the half breadths marked using the graduating batten as before. The initial hewing process was repeated with one important exception. The two sides of the mast now being cut were the athwartship sides. On these sides the contour of the mast was duplicated from the heel to the third quarter. From this point, upward to the head, the hounds were left straight. This flat surface provided solid seating for the cheeks of the hounds.

These operations produced a trimmed stick which, when viewed from the top or sides, revealed surfaces bounded by sharp corners and square cross section at any point.

It now remained to convert these square sections to round, finished sections. This was done by eight-squaring the length of the mast from the heel to the third quarter. To do this 5⁄24ths (why 5⁄24ths? see Fig. 5.1) of each diameter was set off from each side of the middle line on all four sides. These markings were then connected by fair curves along the required length. The material outside of these lines was removed, resulting in an eight-square column, all sides of which were of equal dimension. This is shown in Fig. 5.1, 4th trim. The masthead was now squared to its proper proportions and the tenon cut to suit the cap.

The eight-square stick was now carefully converted to a cylindrical form, the middle line of each surface being retained. The hounds, which had been left four-square, were now hanced to fair into the finished column at the third quarter. The four corners were slightly rounded or relieved with a small chamfer.

Fig. 5.1, 5th trim, illustrates the finished stick.

The preceding description and illustration are intended to portray the general procedure employed in developing a mast made from a single tree. For a detailed description of the process the 3rd edition of ref. 39 is recommended. It is particularly appropriate for masting of the period of this book (1850s), it having been published in 1854.

The furniture required for these masts is described in Chapter 6.

To Make a Trysail Mast

The trysail mast ranged from ½ to ⅓ the diameter of its associated mizzenmast and was of uniform diameter throughout its entire length.

In the absence of specific details it appears probable that the trysail mast started as a four-square stick, out of winding and of the required length. The first step in converting it to round was to strike a middle line along the length of each surface.

Next, a measurement of 5⁄24ths of the width was made on each side of the middle lines and parallel with them. The corners between the new lines were removed forming an eight-square stick.

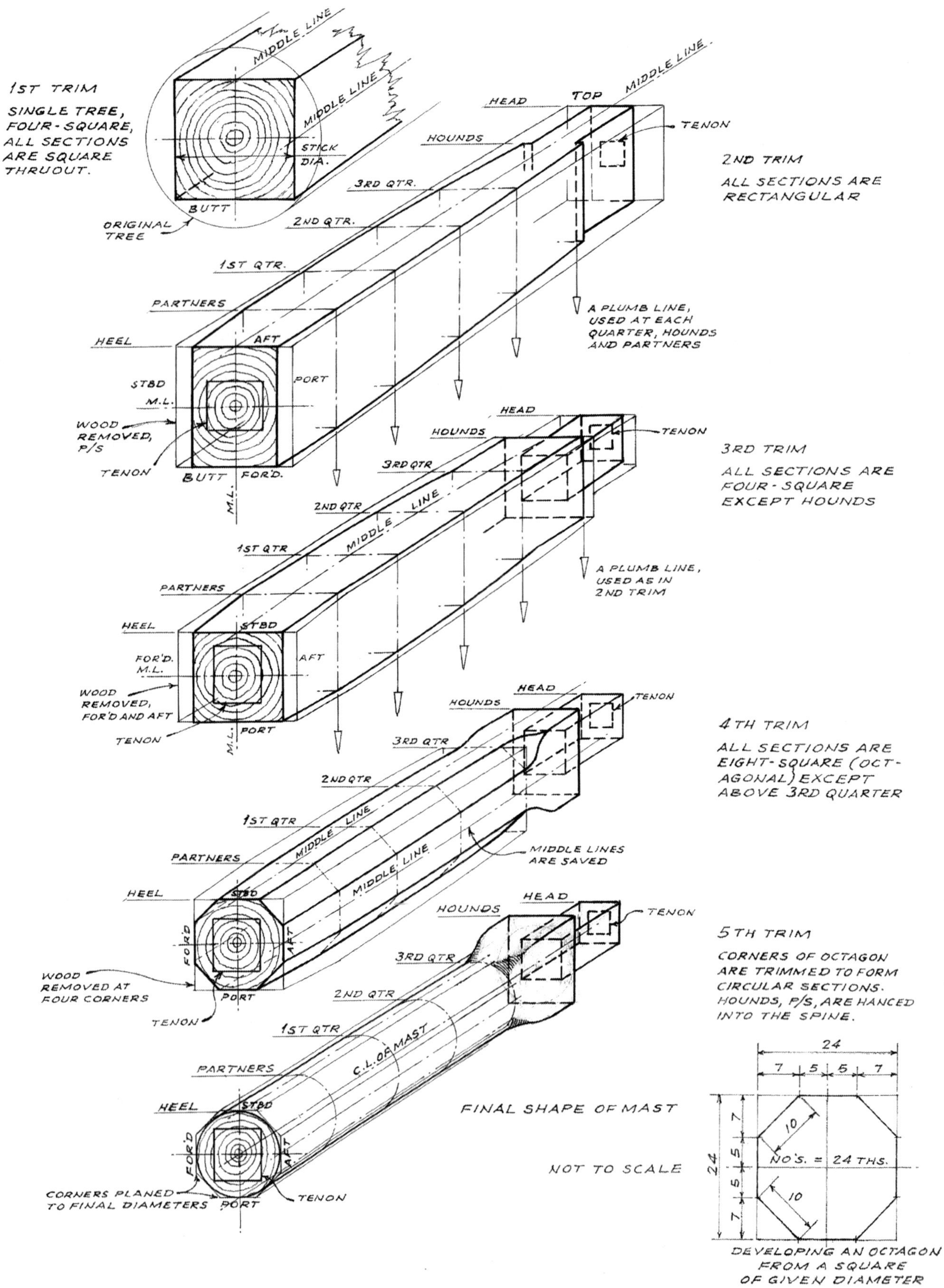

Figure 5.1. Shaping a single tree standing mast or bowsprit.

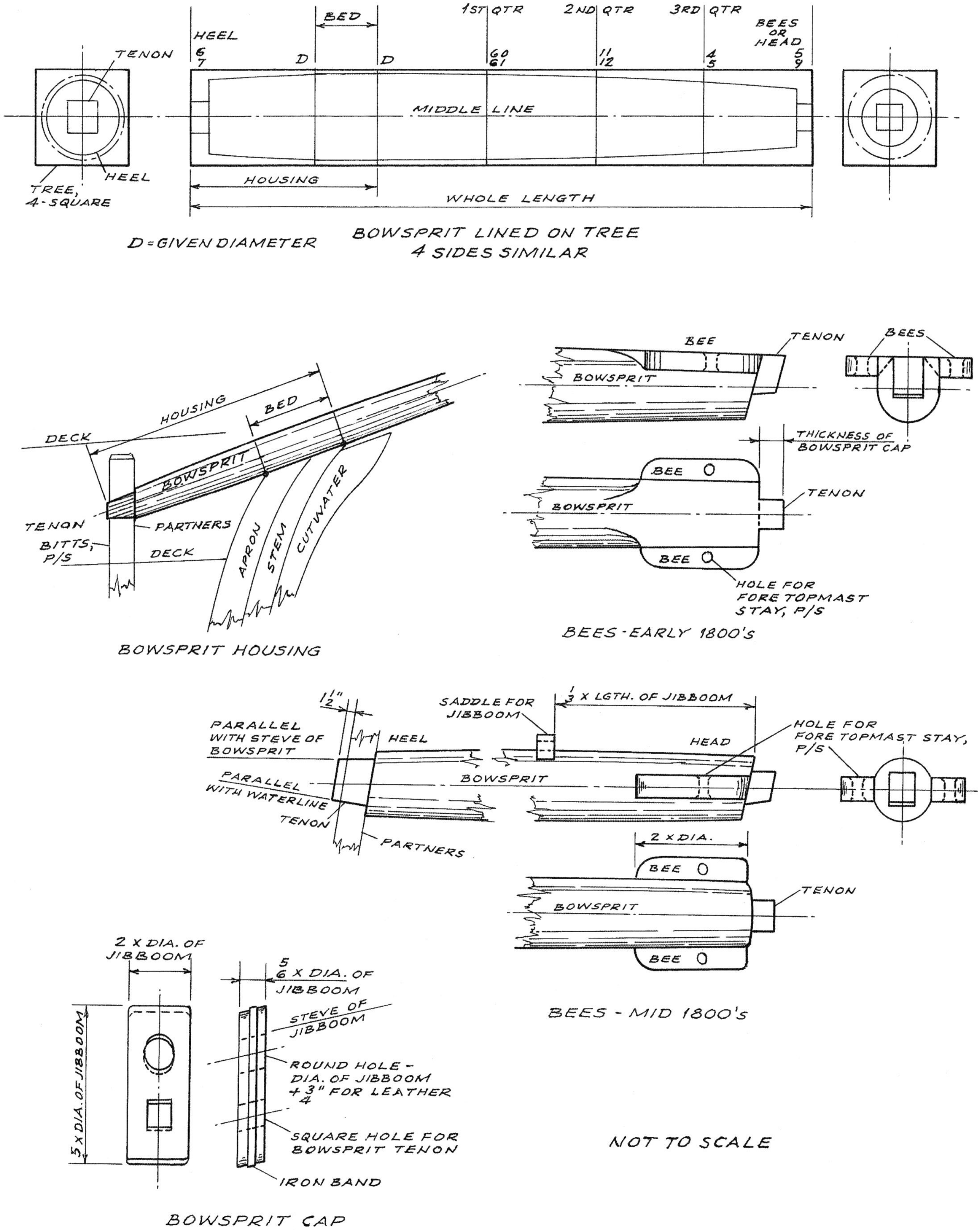

Figure 5.2. Single tree bowsprit and cap.

The eight-square was in turn converted to 16-square, after which it was planed to round along its entire length.

The trysail mast was stepped immediately abaft the mizzenmast and generally rested on a block of hard wood. Its head protruded upward between the lower trestletrees close abaft the crosstree on the aft side of the masthead. Its alignment was made to parallel the mizzenmast.

The trysail mast is shown in Fig. 4.1.

To Make a Single Tree Bowsprit

The procedure for making a single tree bowsprit was essentially the same as that for a single tree mast. A stick of appropriate size was selected and examined for soundness and the presence of imperfections.

If suitable it was hewed to four-square and cut to the desired length. This whole length included a tenon at the head, or outer end, and a tenon at the butt, or partners. In the case of a bowsprit the partners were the bitts that anchored the extreme inboard end of the bowsprit at or near the deck.

Then on the stick a middle line was struck from one end to the other and the location of the bed was lined. The bed was the structure between the forward end of the stemhead and the after edge of the apron. The bowsprit would rest on this surface. The distance between these two locations was transferred to the upper surface of the selected tree. From the after point, or partners, the shape of the bowsprit was lined in. This distance included the length of the heel tenon. The entire distance from the forward point of the bed to the after end of the partners' tenon was called the housing. A detail of this area is shown in Fig. 5.2

Next, from the forward end of the bed to the extreme end of the head tenon, the four quarters are measured off and marked across the tree. This is shown in Fig. 5.2. Each quarter is assigned its proportionate fractional value in reference to the given diameter of the bowsprit.

In measuring the fractional breadth of each quarter it is important to note that the breadth, or diameter, of the bowsprit is constant in way of the bed. This differs from measurements taken for a mast.

The configuration of the bowsprit can now be lined in as a fair curve from the head, or bees, to the forward end of the bed; the line between the forward and after ends of the bed is straight and parallel with the middle line of the bowsprit; from the after end of the bed to the partners the curve is continued.

The bowsprit was identical in shape when viewed from any of the four-square surfaces. Reproducing this shape followed the same procedure as that shown for a single tree mast in Fig. 5.1 except that there was no consideration as was required with the hounds in the mast. By canting the stick and using plumb lines, as shown in that figure, the five trim phases used to convert the mast were applicable to the bowsprit.

The bowsprit, over the years, underwent very few changes and, by this period, had evolved into a very simple, round stick. The bees at the outer end were now mounted as furniture on the extreme end, whereas, at the beginning of the century the end had been left square at its upper sides to augment installation of the bees at the upper eight-squares, as shown in Fig. 5.2.

The second item of bowsprit furniture was the saddle for the jibboom. This was a piece of wood one-half the diameter of the bowsprit in length; one-half the diameter of the jibboom in thickness, and of a height that allowed the jibboom to lie parallel with the middle line of the bowsprit. Its underside was concave to fay with the curve of the bowsprit. It sat across the bowsprit and was located inboard of the bowsprit cap a distance of one-third the length of the jibboom. Its top edge was cupped to cradle the lower side of the jibboom.

The cap was the final item of bowsprit furniture. It was mounted vertically on the forward tenon of the bowsprit. Its length was five times the diameter of the jibboom, its width two times the diameter of the jibboom, and its thickness ⅚ths the diameter of the jibboom. It was a block of elm timber with an oval hole in its upper half to accommodate the jibboom, and a square hole in its lower half to receive the bowsprit tenon. In a fore and aft direction the top and bottom edges of the cap and center lines of the holes were parallel with the steve of the bowsprit. The cap was iron-bound with a band about one-third the thickness of the cap in width. This is illustrated in Fig. 5.2.

If the bowsprit was fitted with an iron cap, the cap was normal to the bowsprit.[33]

Tables covering the sizes and proportions of masts and bowsprits are included in Chapter 4.

6

Made-Masts and Made-Bowsprits

There came a time when the great size of ships made it impracticable to fit them with masts fashioned from a single tree. The first and foremost reason was the difficulty in finding trees with the desired great diameter and necessary straight length. By the period of the 1850s countless trees of suitable size had been harvested over the years and those that remained were located in remote or inaccessible areas. To cut them down and haul them to a logging road or waterway was an expensive undertaking. In some locations such trees simply no longer existed.

The second reason, at least as important as the one just stated, was the liability of such a monstrous timber being unsound in its interior. Certain rot and imperfections could be hiding in a log with no way of being discovered. The finding of such potential imperfections while working a timber could be very costly.

The solution of the problems lay in the practice of uniting two or more trees into a single assembly of the required size. Thus, the made-mast came into being and resulted in the development of the most complex and exacting skill to be found in all of wooden shipbuilding (in my opinion).

To Make a Made-Mast

All sticks that were combined to form a made-mast had at least one straight surface that fayed with a similar surface of another stick with no space between the two surfaces. In a mast composed of only two sticks this was a fairly simple operation. However, in masts composed of many sticks, with many faying surfaces, the task became exacting and room for error became minimal. In timbers which might approach 100 feet in length it becomes obvious that skill and accuracy were paramount.

In comparison, the Table of Offsets, which contains the dimensions that control the moulded shape of a vessel's hull, accuracy is held to within ⅛th of an inch for any dimension.

There is no attempt in this book to cover in minute detail the complete procedure for fabricating a made-mast. The subject could fill a volume in its own right. However, the basic details which cover the general concept of mast construction are discussed.

When sticks were prepared to fay, coaks about one inch high were left to stand above the plain, which is the flat surface that is left between coaks and is the finished surface of the stick.

In the stick that was to fay with the coaked stick, mortises corresponding in size, depth and location were sunk below the surface. When the two sticks were placed together there was no space between them, and it was impossible for one stick to shift in reference to the other.

This procedure was, in general, followed for all masts, some of which could contain as many as 16 individual sticks.

The aggregate combination of sticks was fashioned to make a mast of given diameter. The sticks each had a name which was based on its position in the final assembly.

The principal sticks were the spindle, or core, which was the central stick about which the mast was built. This stick was positioned with the butt end upward.[39] In some cases it extended for the full length of the mast, from the heel to the head; in other cases, if short, its length was extended by the installation of lengthening pieces.

Next in importance were the side trees which were placed on the athwartship sides of the spindle, butt end downward, and sized to fill out the desired diameter of the mast in the athwartship direction.

Following the side trees were the fishes. These were placed on the forward and aft sides of the spindle and side tree assembly and were sized to complete the diameter of the mast. They were positioned with the butt end downward.

In addition to these principal sticks there were minor lengths of timber called fillings which were shaped to fill in any edges around the circumference of the mast and thus complete the assembly.

The sticks in an assembly were bolted together as a unit in order to prepare the assembly for final hooping with substantial iron bands or hoops.

The final pieces to be added to the above assembly before being hooped were the cheeks. These were long structures, one on each side of the mast, which extended from above the second quarter, at their lower end, to the top of the hounds. The inside surface of the cheeks was hollowed out to fay with the surface of the mast assembly.

The lower end, or tail, was in width about one-half the diameter of the mast. The upper end was as wide as the full diameter of the mast and, at the hounds, converted into a square configuration, its purpose being to support the lower trestletrees. From this upper end to the lower end of the tail the cheek was gently hanced from square to half-round and uniformly reduced in thickness until, at the lower end, the tail was bearded to the surface of the mast. See Figures 6.1 through 6.4.

The upper surface of the cheeks, at the top of the hounds, was cut at an angle which corresponded to the rake of each mast so that the trestletrees, when installed, would parallel the ship's waterline.

With the cheeks bolted in position the mast could now be hooped. The locations of hoops were marked on the bolted assembly and the girths taken around the mast at the appropriate stations. Hoops varied in width and thickness depending upon the size of the mast to be hooped. In general these iron bands were about four inches wide and ½ inch thick.

In most masts there were two types of hoops required. The first, and most common hoop, was formed to a true circle to embrace a specific location on the mast. The second type of hoop was known as a mould hoop. These were required in locations in which the true diameter of the mast was infringed upon, such as in way of the cheeks.

In these locations a mould was made which included the sectional shape of the cheek at a specific location. A special mould was required for each hoop to accommodate the variable dimensions of the cheek. The hoop was then shaped to its proper configuration while cold.

The leading edge of each hoop was chamfered on the inside to avoid digging into the mast surface when being driven into its final position. The usual spacing between hoops was about three feet.

In preparation for being driven the hoops were heated until nearly red, then driven to their particular stations by long, round rods which were flattened at the foot. These rods were called pokers. The foot was placed against the mast, one on each side, and racked back and forth as the hoops were pounded to advance them along the mast to their final location. The mast previously had been liberally greased with tallow to prevent burning the wood. When at their assigned station the hoops were doused with cold water to hasten shrinkage and chill the hoop.

With masts being built by the thousands it is not difficult to accept the fact that, at some time, a mast maker would require trees of a size that he did not have and could not acquire. When this did happen designs for the construction of made-masts were originated based on the sizes of trees that were available. Examples of such specimens are shown in Figures 6.1 through 6.4. These figures illustrate the form of the individual pieces only and do not include any details such as coaking, hooping, etc.[39]

A Common Made-Mast

Fig. 6.1 is referred to as a common made-mast. It consists of a spindle made of two sticks (a) which extend from the head of the mast to a point below the partners; two side trees (b) which extend from the heel to the third quarter; two cheeks (c) extending from the second quarter to the head of the mast at the cap; and two fishes (d) that extend from the heel to the third quarter; a total of eight sticks.

The configuration of all these pieces is plain and very basic. It might be assumed that the spindle (a) is placed with the butt end upward and the side trees (b) are placed butt end downward. The original text,[39] however, makes no note of this.

This mast combines simplicity of construction and timbers of a size that was readily available or easily fashioned.

This mast is sometimes constructed with the spindle (a) made from a single stick.[39]

A Top and Butt Mast

Fig. 6.2 depicts a made-mast which was known as a top and butt mast and was generally made for masts between 24 and 32 inches in diameter. It was made of one heel piece (a), which extended from the heel of the mast to the mid-height of the hounds, and a head piece (b), which extended from the top of the masthead to the heel, if a stick of sufficient length was available, or to a point below the partners if the piece was short. In the latter case a lengthening piece (c) was added with a scarph five or six feet in length. The scarph was always below, and clear of, the partners. In addition there were two cheeks (d) which extended from the top of the masthead down to the second quarter where they were bearded into the mast.

The construction of this mast was peculiar in that the common faying surfaces of pieces (a) and (b) were not on the centerline of the mast. Instead, the lower end of the heel piece (a) occupied more than half the area of the heel. From this point the faying surface of the heel piece angled up on a diagonal, crossing the centerline of the mast. At its upper extremity it was about two inches thick where it abutted piece (b). The upper end of the head piece (b) was large and the lower end small. Between them they filled the fore and aft diameter of the mast.

In this construction the butt end of the tree used for the heel piece (a) was installed downward; the butt end of the tree for the head piece was installed upward. This accounts for the designation top and butt mast.

The final pieces of the construction were the two fishes (e) which extended from the heel of the mast to the third quarter. The inner surfaces of the cheeks were hollowed to fit the fishes.

This mast was composed of seven pieces.[39]

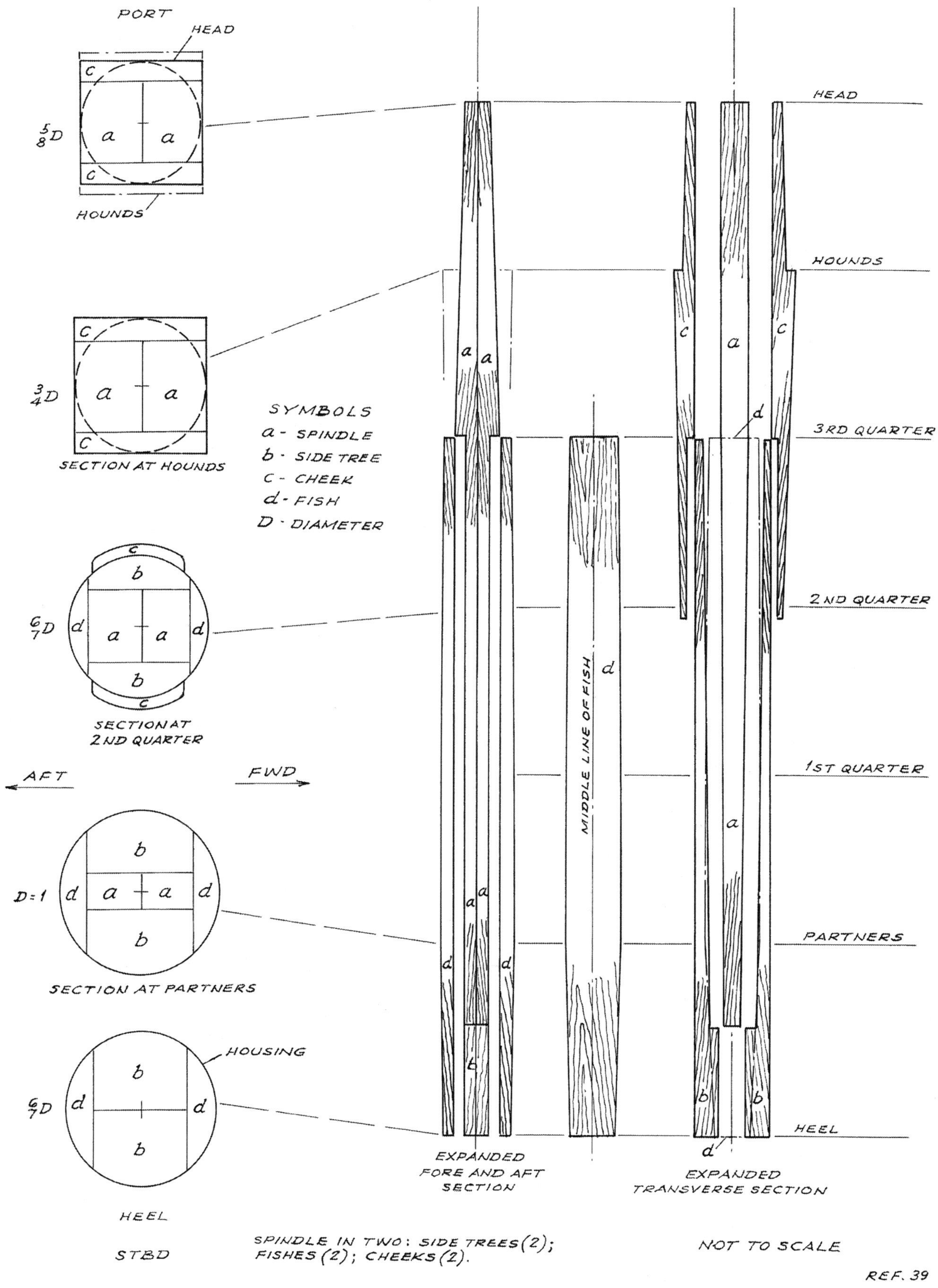

Figure 6.1. A common made-mast.

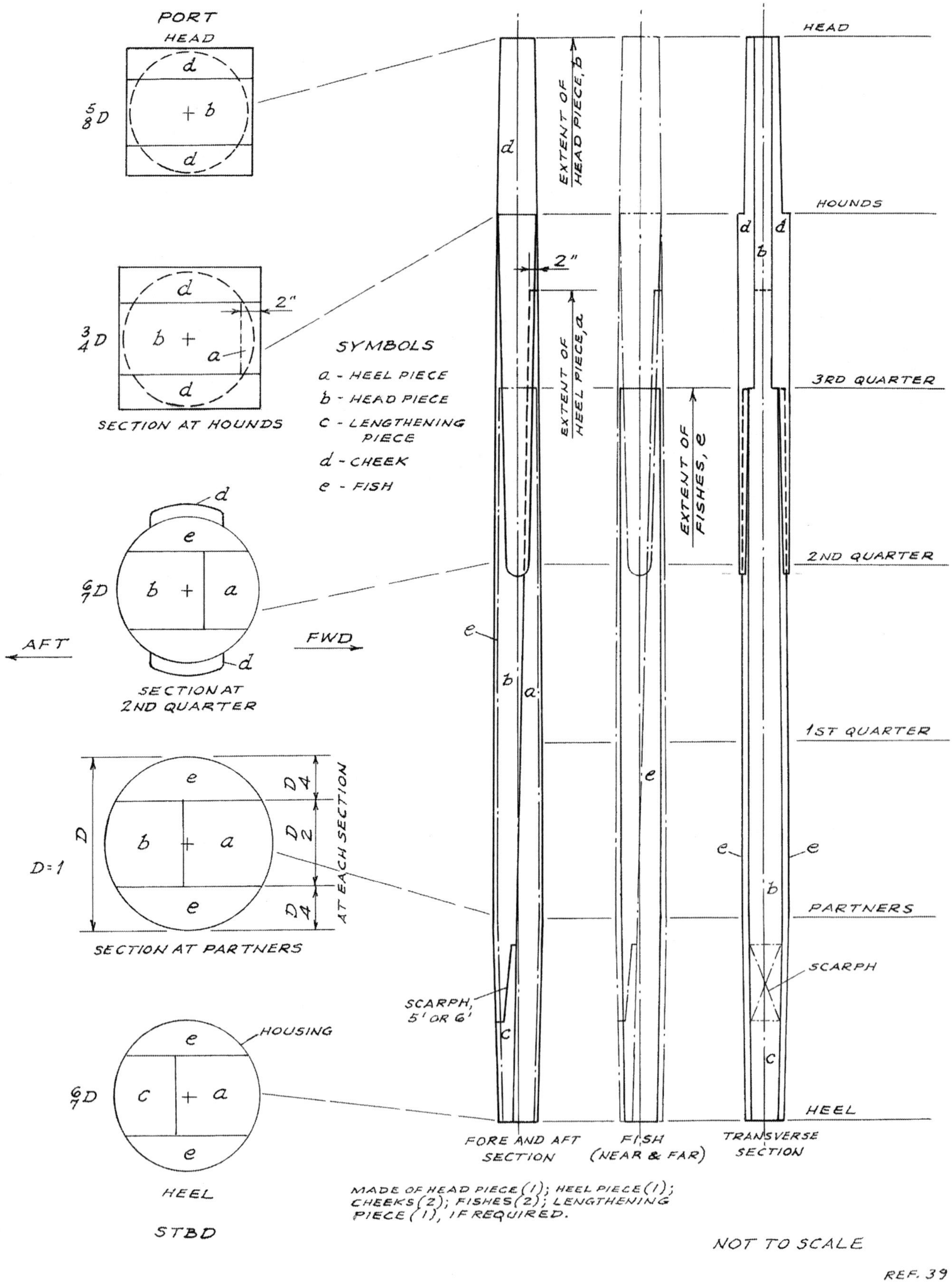

Figure 6.2. A top and butt mast.

A Mast with a Short Spindle

Fig. 6.3 is a mast having a short spindle, designed for use in cases where trees of long dimension were not available. It was composed of two side trees (a) which extended from a scarphed end below the partners, to the lower end of the cheeks (c), above the third quarter; a one-piece spindle (b), the lower end of which terminated below the second quarter and whose upper end reached the top of the masthead; two cheeks (c) which extended down from the top of the masthead to about the third quarter; two fishes (d) which reach from the heel to the lower end of the cheeks; and two fillings (e) inserted under the cheeks if the side trees cannot hold their full dimension under the cheeks in way of the hounds. Finally, there were two lengthening pieces (f) which were scarphed to the lower ends of the side trees to complete their lengths to the heel, if required.

This mast was composed of 11 pieces.[39]

A Mast Made of Small Trees

Fig. 6.4 shows a mast that is formed of small trees and is assembled without a spindle. The diameter of the mast is attained by fitting four sticks of identical shape, each filling one quarter of the total mast area. The four sticks were fayed to each other and formed a solid mass.

This mast was designed for the use of small, short sticks. In addition to the use of smaller trees it was convenient when conveying them from one site to another.

It was conceded that this mast, having no spindle, would be less rigid than an ordinary common made-mast but that it would be capable of bearing all the stresses to which masts were subject.

The four principal trees (a) were each made up of two or three scarphed lengths. The scarphs (c) were about six feet in length and were located so that there was ample shift between them in an individual stick, and that no scarphs would oppose each other. Each of the individual sticks extended the full length of the mast from the heel to the top of the masthead.

Two cheeks (b) extended on the sides from the top of the masthead to a point below the second quarter. These were hollowed on the inside to allow the tail of the cheeks to be bearded into the mast.

An abutment was formed in the trees (a) above the third quarter to support the cheeks.

Figs. 6.1 through 6.4 illustrate the wide range of details which were combined to construct a made-mast. The basic design, that is a spindle, or core, acting as a central column, augmented by a surrounding collection of lesser trees or sticks, found expression in many ways. Much of mast design was governed by the availability of desired timber.

The details shown in the aforementioned figures are generic and cannot be assigned to any specific vessels. Apparently, for any given vessel, a mast of any conventional construction was acceptable if its dimensions agreed with those specified by the ship's designer. There is very little specific information available on this subject.[39]

Many variations of made-mast structure existed. Some were attributable to known individuals but most remain anonymous. What detailed information is known is of British origin. Unfortunately, American innovation, if any, and practice, remain unknown quantities.

Fig. 6.5 shows illustrations of cross sections of made-mast construction ranging from masts made of two sticks to those that are composed of as many as 16 sticks.

Detail A shows two trees, each making up one-half of the diameter of the mast. The peculiarity of the construction lies in the fact that the two trees are not divided on the centerline of the mast. In this instance the heelpiece covers about two-thirds of the butt, or heel, of the mast while the head piece covers the opposite two-thirds of the upper end. The two pieces fay and cross the centerline of the mast at about the mid-point of their length. This construction is also shown in Fig. 6.2, the top and butt mast.

Details B and C show masts made of three and four trees, both of which were covered by British patent by a man named Hillman.

Detail D shows a mast with an octagonal spindle surrounded by eight identical pieces fayed to it. It extended the full length of the mast from heel to masthead. This design was covered by British patent by a man named Ferguson.

Details E, F and G show typical variations of mast structure that may have been devised to accommodate material that was available.

Finally, Detail H shows a mast constructed in strict fasces fashion. It was a bundle of 16 sticks, all of similar size, bound into a single unit. This was a concept conceived by Sir Robert Seppings, a British mind who was very active in mast-making at this period.

Chapeling a Mast

The chapeling of masts is an operation which receives scant, if any, attention in books dealing with shipbuilding. The derivation of the word does not appear in dictionaries or the few references that mention it at all. Short descriptions of the procedure are to be found in three nautical books which appear herein as references 35, 45 and 51. The contents of each book are discussed separately.

In Ref. 35, *The Mariner's Dictionary* by Gersham Bradford, the author described chapels as "grooves in a built-up mast. The appearance of the vessel was enhanced in the sailing ship days by painting the chapels in contrast with the general color of the mast."

In Ref. 45, Capt. E. Armitage McCann, in the February 1930 issue of his magazine *The Ship Modeler, Official Journal of the Ship Model Maker's Club*, wrote that "between the iron

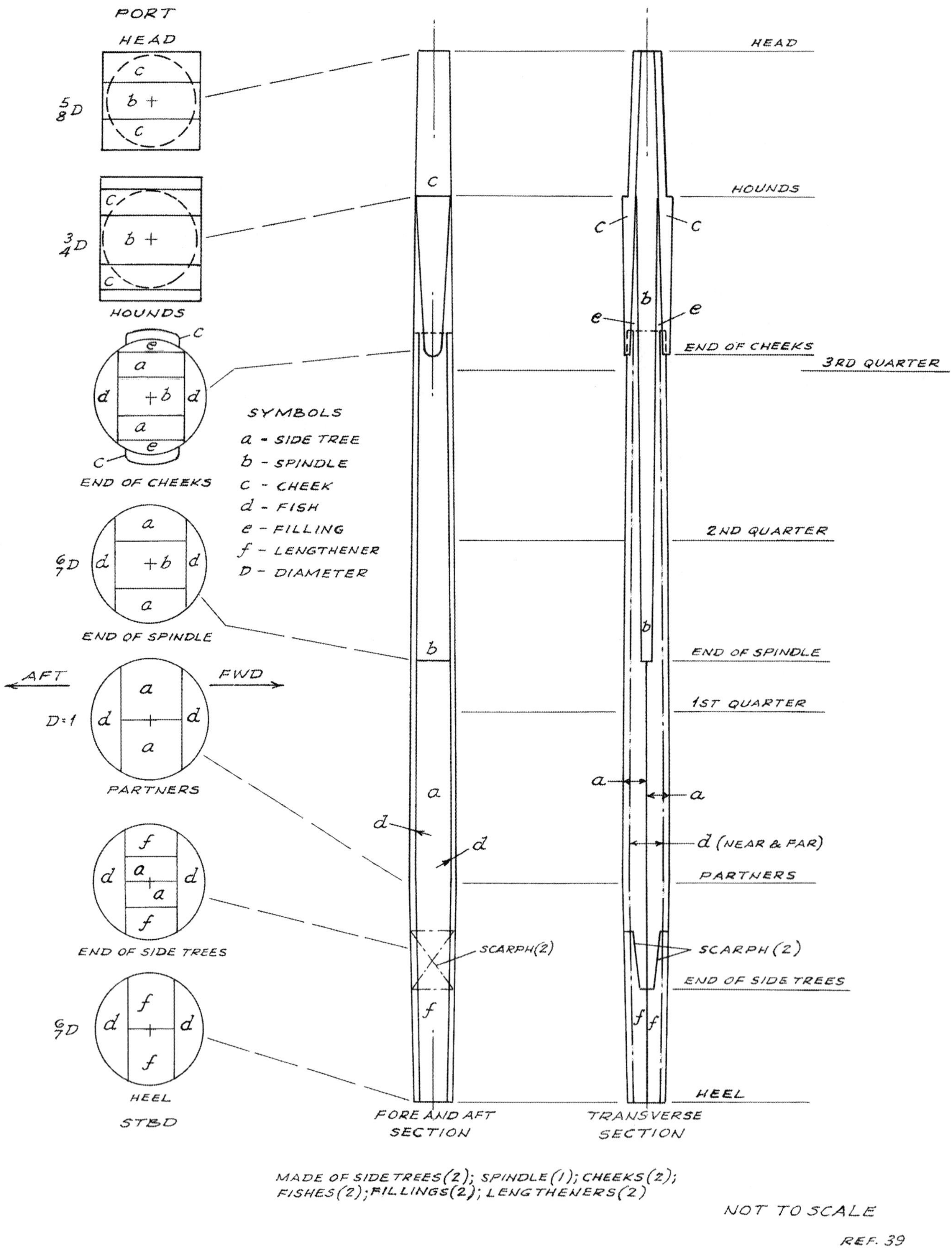

Figure 6.3. A mast with a short spindle.

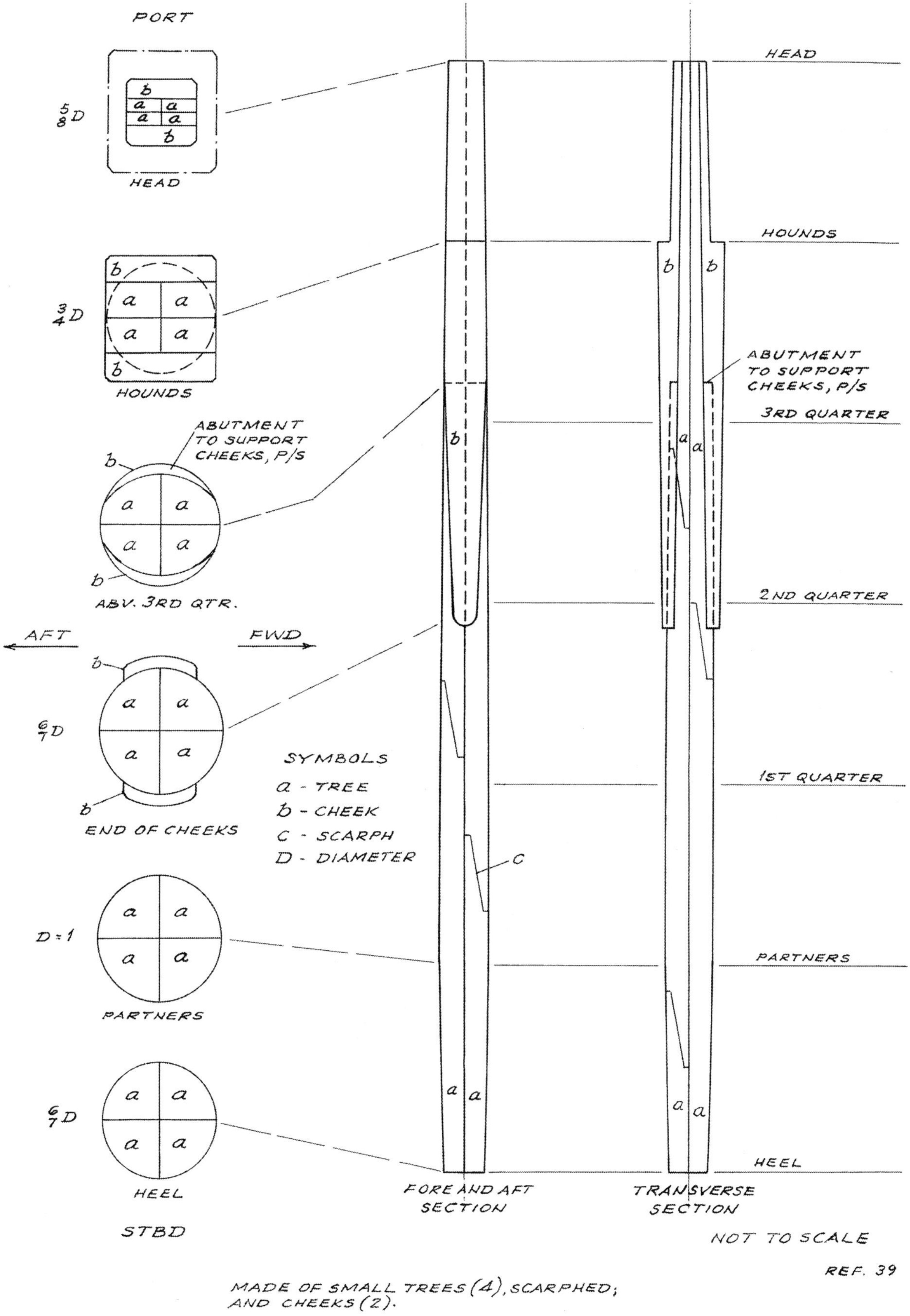

Figure 6.4. A mast made of small trees (scarphed).

mast bands of lower masts that were 'made'" (built-up) prism shaped notches, called "chapels," were cut. They were purely ornamental and were painted white. The text was accompanied by the illustration Fig. 6.6, Detail 1.

In the March 1930 issue of the same magazine he wrote, "Chapeling was sometimes, in the more roughly made masts, the space between fillers under the mast bands." This text was accompanied by illustration Fig. 6.6, Detail 2.

Nowhere is there an explanation as to why chapeling was required. My conjecture is that the long grooves which resulted from beveling the outer corners of the individual sticks, relieved any sharp, vulnerable, protruding edges, Fig. 6.6, Detail 3.

In Ref. 51, *The Oxford Companion to Ships and the Sea*, edited by Peter Kemp, chapels were described as "the grooves in a built-up, or 'made' wooden mast, in which several pieces of timber are used to fashion it.... The chapels occurred where the various pieces of which the mast was made were joined together."

All the masting detail in this book, except the chapeling details in Ref. 45, which are wholly American, has been derived from British sources, Refs. 33, 39 and 47. This is unfortunate for a book purporting to be about masting of the American merchant fleet of the 1850s. However, were it not for such information there would be little or no knowledge from which to refer.

What is known about American masting does form a considerable portion of the total content of this volume. Masting tables in the Appendix contain the spar dimensions of many American vessels and also contain some specific detail concerning their masting.

For instance, many descriptions record that specific vessels were fitted with single tree or made-masts. More importantly, a few ships were listed with limited, but specific, information, as follows:

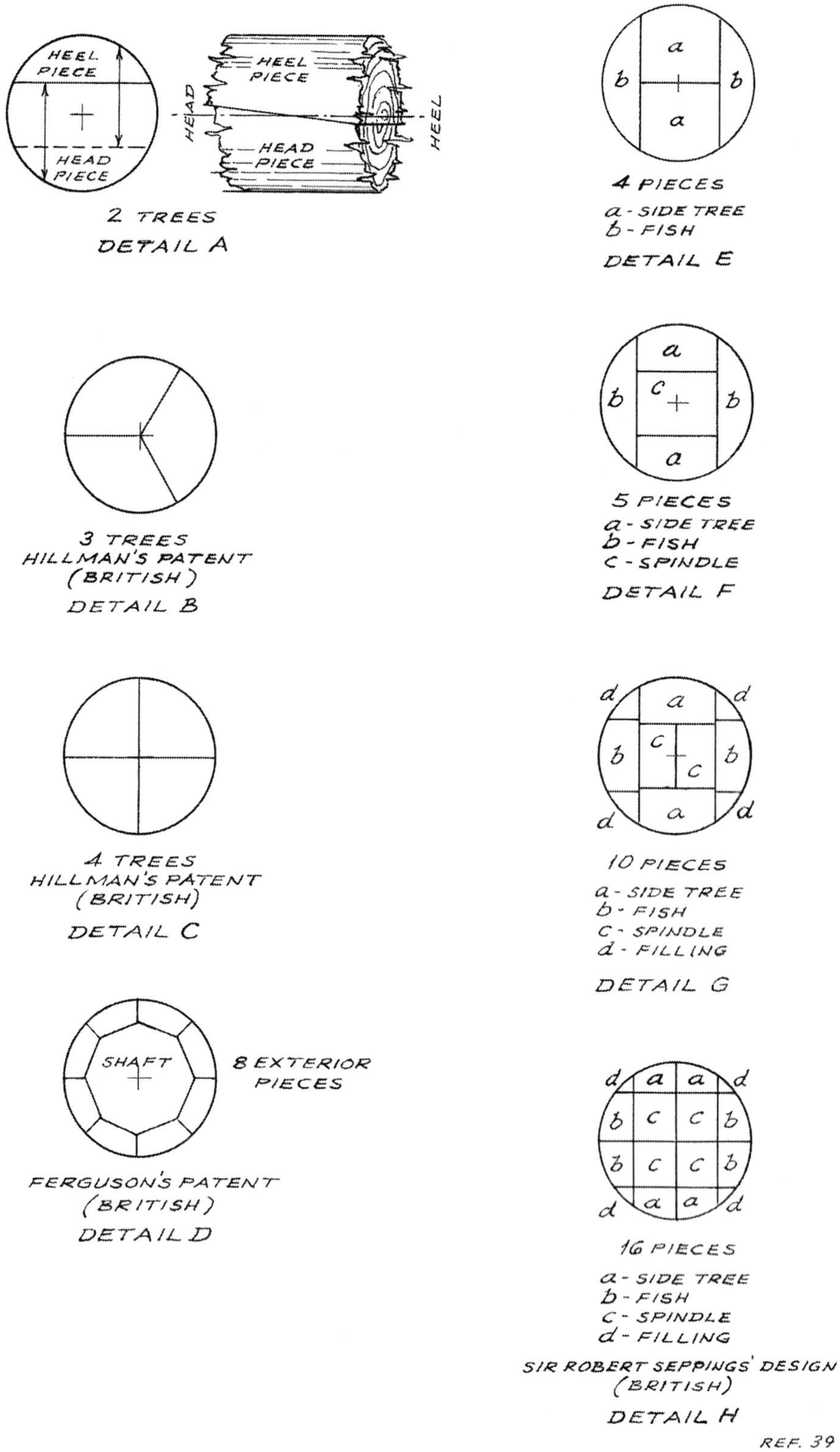

Figure 6.5. Sections through made-masts.

1851 *Flying Fish*,[1] "fore and main lower masts are made, and are fished on every square and hooped. Fishes extend two feet below lower deck partners."

1851 *Hoogly*,[1] "fore and main lower masts are made, each in five pieces, hooped."

1851 *Invincible*,[6] "lower masts are made of five pieces and filled in under the hoops."

1850 *Stag Hound*[1] and *Staffordshire*,[1] "fore and main lower masts are made and fished on the sides."

1852 *Bald Eagle*,[1] "All lower masts are made, the outside

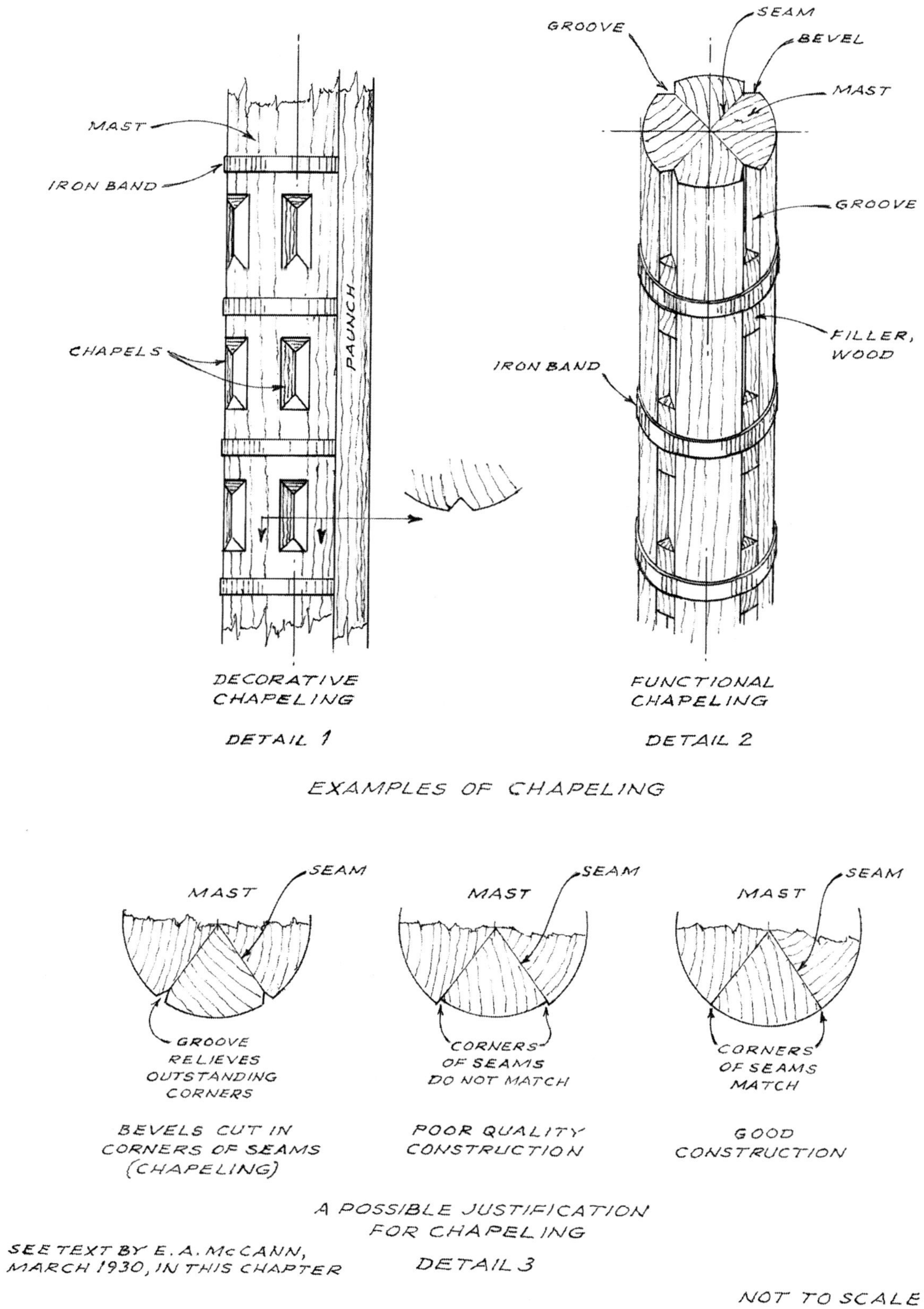

Figure 6.6. Chapeling a mast.

pieces extending the whole length from steps to the heads, doweled together, bolted and hooped over all."

1853 *Mystery*,[1] "fore and main lower masts are made ... each of 14 pieces, bolted and hooped."

1855 *Donald McKay*,[1] "Lower masts, and bowsprit, are of pitch pine, and are hooped with iron. Hoops on fore and main mast are four inches wide by ⅝ inch thick, 30 and 31 in number. Mizzenmast has 28 hoops."

This is the sum total of information on mast construction for the 152 vessels included in this book. For a subject covered in such depth as the American sailing ship this phase has been sorely neglected.

Furniture for Standing Masts

The masthead of a standing mast was encrusted with a group of individual pieces known collectively as furniture and specifically as a top. The top was a semi-circular platform projecting around the lower extremity of the masthead supported by the cheeks of the mast. Its purpose was to spread the lower ends of the topmast shrouds and thus increase support of the upper masts.

Merchant tops were made as small as practicable to reduce chafing of the topsails against the rim of the top. Their breadth was equal to one-half the moulded beam of the vessel; their length was one-half of their breadth.

(At this period the great spread of topmast rigging, as achieved with excessively broad tops, was not considered as effective in supporting the upper masts as a good stout standing backstay.)

The aft end was straight and athwartship. The forward portion was a semi-circle struck with its center on the after rim at the middle line of the ship.

Five-twelfths of the length of the top stood aft of the centerline of the mast; seven-twelfths stood forward of that location.

When fitted on the mast the top and the upper surface of the cheeks were always at an angle which allowed the top to sit parallel with the ship's waterline.

In addition to the top there was a fitting known as the cap. This was located at the top of the doublings and served to constrain the lower mast and the topmast in a fixed relationship. The cap was of an oval configuration and was generally made of one piece.

The top is illustrated in Fig. 6.7 and was composed of the following pieces:

Bibbs (a) are elm and are fitted one on each side of the mast. They fay with the underside of the trestletrees and the forward edge of the cheeks. They are four-fifths of the hounds in length; in breadth they reach forward the diameter of the topmast; in thickness they are half that of the trestletrees but never less than three inches. They fay with the cheeks with a hook scarph having the hook point aft to support the trestletrees. Their purpose was to counteract the overturning load of the upper masts. They were bolted edgeways through the cheeks. Their upper edge sloped forward to suit the rake of the masts.

Trestletrees (b) are longitudinal timbers which rest on the cheeks and extend the full length of the top. Their purpose is to support the upper mast assemblies. Their depth is one-half the given diameter of the mast; the breadth is two-thirds of the depth. The forward ends are rounded down and the aft ends trimmed to a long snape. They are bolted to the masthead.

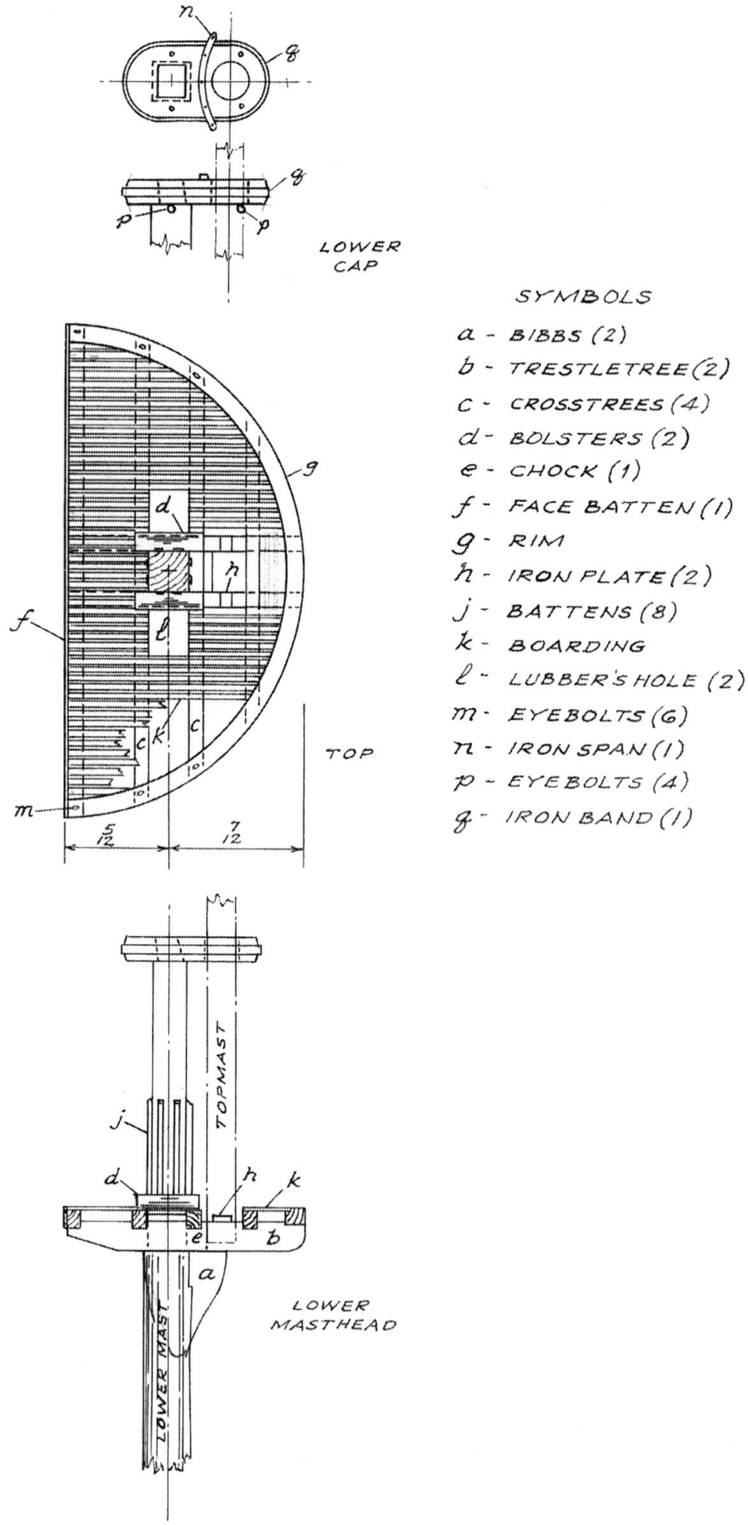

Figure 6.7. Furniture for a lower mast.

Crosstrees (c) support the structure of the top. One crosstree is fayed with each of the forward and aft sides of the masthead; one frames the after edge of the top; a shorter one frames the forward edge of the opening for the topmast. They are symmetrical about the middle line of the top. The breadth of the crosstrees equals the thickness of the trestletrees; their depth is two-thirds the thickness of the trestletrees. When assembled they are let into the trestletrees.

Bolsters (d) are short pieces of fir fayed upon the trestletrees and against the sides of the masthead. Their ends rest on the crosstrees which embrace the mast and are bolted thereto. They are of thickness about 1½ inches greater than the trestletrees and are trimmed down as quarter-rounds. They prevent the shrouds from being chafed by motion of the mast.

Chock (e) is a spacer on the fore side of the mast which is snugged between the trestletrees. It is as deep as the trestletrees and, in thickness, the same as the distance between the two holes in the cap. Its purpose is to keep the topmast in the same vertical alignment as the standing mast.

Face batten (f) is a thin board which spans the breadth of the after edge of the top and frames the bitter ends of the boards which are the platform.

Rim (g) is a wooden border about eight inches wide and one inch thick which frames the circular portion of the top. It does not span the square after end.

An iron plate (h) is let into the top surface of each trestletree in way of the heel of the topmast to provide a bearing surface for the topmast fid.

Chafing battens (j) on the masthead, are oak or fir, and are, in length, about ⅖ the length of the masthead. The upper ends are snaped sharp into the masthead. They range in width from two to four inches, and from two to three on each side depending on the size of the masthead. They are about 1½ inches thick. Their purpose is to prevent the eyes of the rigging from chafing against the masthead.

Individual boards or battens (k) compose the surface of the top. They are 1½ inches thick and about three inches wide, placed with a space between. This arrangement was beneficial in two ways. First, the structure was relatively light in weight, and second, its grated construction did not hold the wind.

Lubber's hole (1) allowed access to the top from below without having to go over the rim by means of the futtock shrouds.

Eyebolts (m) are installed through the rim at the ends of the three aftermost crosstrees to anchor the topmast shrouds.

The Lower Cap is generally made of elm due to its curly grain's resistance to splitting. Its depth is ⅚ of the diameter of the topmast; its breadth is twice the diameter of the topmast. Its length is approximately as follows: one thickness of the cap aft of the square hole, plus the dimension of the masthead, plus the thickness of the chock (e), plus the diameter of the topmast, plus one inch, plus ⅔ the depth of the cap.

An iron span (n) is fixed athwartship on the top of the cap between the two holes. It has an eye in each end to receive the lower lifts. In width it is ⅛ the depth of the cap; its thickness ranges between ⅜ inch and ⅝ inch, varying with the size of the ship. It is secured with three bolts.

Four eyebolts (p) are fitted to the underside of the cap; two are abreast of the centerline of the square hole; two are abreast of the forward half of the round hole. These receive the top-rope pendants.

An iron band(q) ⅓ the depth of the cap and about ½ inch thick surrounds the periphery of the cap.

The cap is illustrated in Fig. 6.7

Tops were sized as follows[33]:

Breadth of main top = ⅔ to ½ the moulded beam of ship.

Length of main top = ⅔ to ½ the breadth of top.

Breadth of fore top = ⁹⁄₁₀ of main top.

Breadth of miz. top = ⅘ of fore top.

Some known details of tops for ships included in this book follow.

1851 *Challenge*,[1] Tops are solid (naval fashion) and fit close to the eyes of the lower rigging. Fore top is 16½ feet wide; main top 17 feet wide; mizzen top 13½ feet wide.

1851 *Staffordshire*,[1] Tops are boarded solid (naval fashion) and are close to the eyes of the rigging.

1851 *Swordfish*[1] The fore top is 14 feet wide; main top is 14½ feet wide; mizzen top is 11¼ feet wide.

1852 *Flying Dutchman*,[1] The fore top is 14½ feet wide; main top is 15 feet wide; mizzen top is 11½ feet wide.

1853 *Empress of the* Sea,[1] The fore and main tops are 17 feet wide; mizzen top is 13 feet wide.

1853 *Young America*,[2] The fore top is 17 feet, nine inches wide; main top is 18 feet, six inches wide; mizzen top is 14 feet, six inches wide.

To Make a Made-Bowsprit

Unlike the made-mast, which was constructed in accordance with many designs and varying component parts, the made-bowsprit, in the 1850s, was constructed in the same detailed fashion that had prevailed for at least 50 years.

No exact date can be assigned to the origin of the design, and there is no record of whether the ultimate bowsprit was merely the result of experiment and experience. However, this design and method of construction is explained in Steel's *Elements of Mastmaking, Sailmaking and Rigging*, published in 1794 (Ref. 47), and repeated, in essence, in Fincham's *Masting Ships and Mast Making*, published in 1854 (Ref. 39), originally published in 1829. General details are shown herein, in Fig. 6.8.

Two trees of the required length, which included the heel tenon and the tenon at the outer end of the bowsprit, are first inspected for the usual defects, such as decay, unsound knots, shakes, etc. If found suitable both are trimmed four-

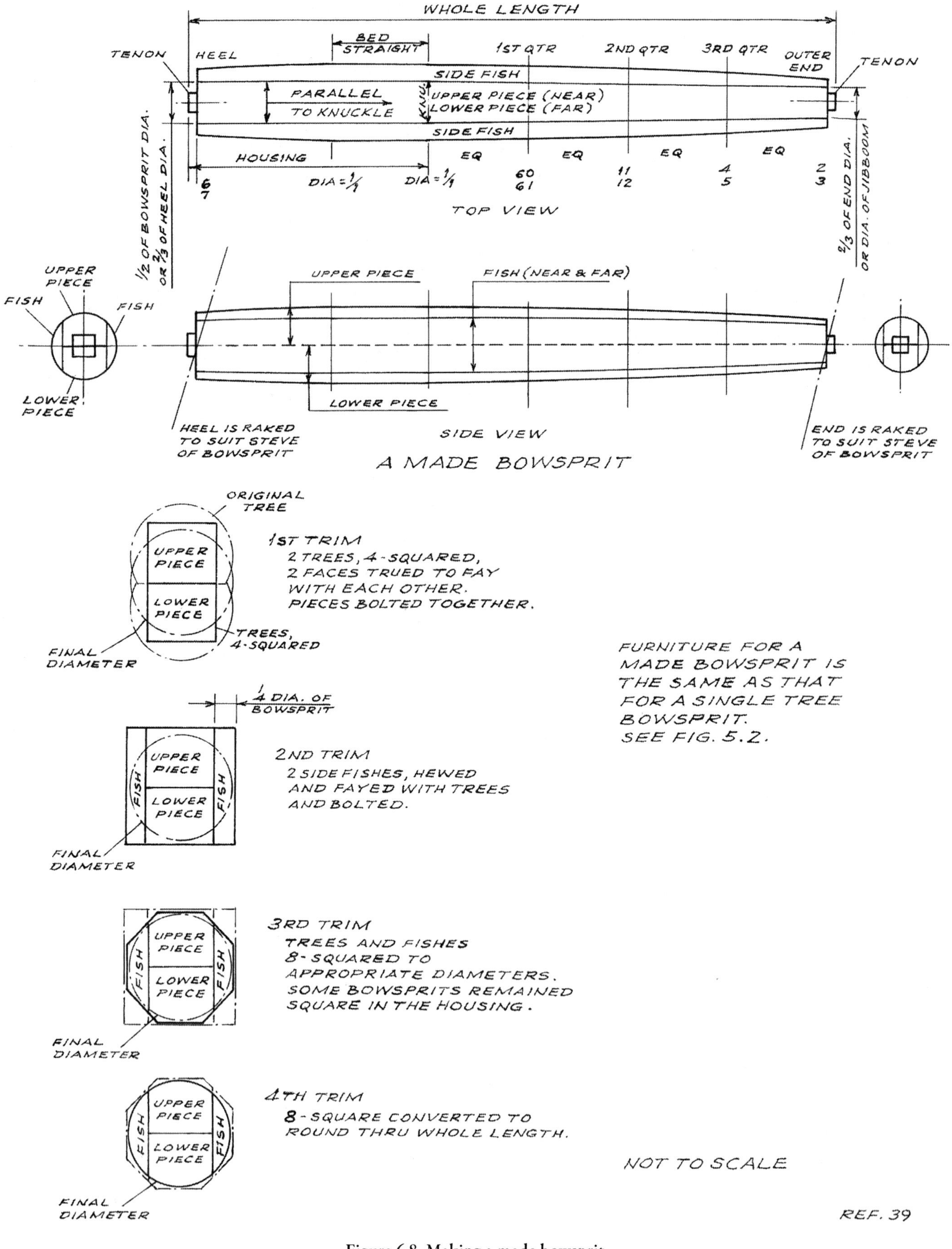

Figure 6.8. Making a made bowsprit.

square. Their combined dimension, one on the other, must be great enough to provide the given diameter of the bowsprit.

The two sides that will fay are trued, coaks are raised on one, and matching mortises sunk in the other. When suitably fitted they are bolted together. The combined upper and lower pieces comprise the given diameter of the bowsprit. The fayed surfaces form the horizontal centerline plane of the finished structure.

The butt end of the trees forms the heel of the bowsprit; the top of the trees forms the head or outer end.

This assembly is laid horizontally, the upper and lower pieces in their assigned position, and a middle line is struck on the top piece from one end to the other. Across the breadth of the top piece are lined the butt end; the limits of the bed, which is the given diameter; the quarters and the outer end. The tenons terminate each end.

At the heel, or butt end, a width ⅔ of the heel diameter, or ½ the bowsprit diameter, is centered across the stick. This is repeated at the forward end of the housing. Parallel lines are drawn between these points. At the outer end, or head, of the trees, a line ⅔ of the end diameter of the jibboom is lined off in a similar manner. The points at the housing and the points at the outer end, on each side of the middle line, are connected. This results in a straight taper between the housing and the outer end. The material outside of these lines is sawed away. The sawed surfaces are rendered smooth and true.

Next, to attain the athwartship diameter of the bowsprit, two side-fishes the length of the bowsprit are fashioned. According to Steel,[47] "Side-fishes were commonly sawed out of one tree straight down the middle and ¼th the diameter off on each side for the thickness."

These side-fishes extend the full length of the bowsprit and are bolted in place. The lines for the ends, quarters, etc. are struck around the diameter of the assembly which is then converted eight-squares and 16-squares to its final, round configuration. Sometimes the housing was left square.

The final array is hooped with iron, one hoop at the heel, one at the head and, in between, at a spacing of about three feet between centers in the manner of masts.

Furniture for the made-bowsprit was the same as that for the single tree bowsprit. This is shown in Fig. 5.2.

The procedure of lining, canting and plumbing the individual pieces is the same as with single tree masts. See Fig. 5.1.

Modifications to the Standing Mast

By the late 1700s the design, development and fabrication of the standing mast was regarded as the epitome of its progress over centuries of use.

Little of the knowledge that was gained over this long period was based on theory. Most was absorbed through constant observation and comparison of one model with another. Weaknesses were eliminated; strengths were incorporated. Proportions were adjusted and modified until, ·in the end, the most efficient compromise was selected.

The ultimate, general design of the masts proved to be very efficient for the purpose intended, however, in the overall picture there were some rather perverse results.

All structures, regardless of their strength and efficiency will, if put to the ultimate test, divulge the weakest part. So it was with masts, bowsprits and yards.

In a mast the failure most likely to occur was the wringing of the masthead where the cross-sectional area was suddenly reduced in reference to the body of the mast.

Bowsprits were most likely to wring between the gammoning and the cap.

Yards generally failed in way of the yardarm where the grain of the wood exited the substance of the tapering yardarm.

These shortcomings were known and accepted as the least serious of the accidents that might happen.

Proportion is the paramount rule in shipbuilding and masts were no exception. Countless tables were compiled and lists published containing every conceivable proportion of one part to another. Consequently, the top-hamper of one vessel greatly resembled that of any other similar vessel. Some of these tables and lists appear in Chapter 4.

The 1850s saw the first and only deviation from these standards. The cause was the routine appearance of 2000 ton vessels which brought with them proportionate increases in the sizes of masts and yards. Of particular concern was the great increase in the area of the single topsail.

Shipping minimal crews merchant ships were having more and more difficulty in handling this sail under adverse conditions. The problem was one that had been increasing for a long period and many minds ashore were busy trying to discover a solution, not only in the United States but also in other maritime nations. Dividing the single topsail into two smaller parts became the solution to the problem.

Capt. Robert Bennett Forbes, of Boston, had been working on this for some years, having first fitted them in the bark *Edith*, in 1844; the ship *Massachusetts*, in 1845; and the ship *Samoset*, in 1847. By the 1850s his rig was well established.

The relevance of this double topsail rig to the subject of masts as discussed in this book lies in the fact that the length of the lower mastheads was increased appreciably to accommodate this new rig. This modification was the first change in masting in more than 50 years. The change is shown in Fig. 6.9 under Forbes' rig.

With Forbes' rig the only change from a conventional mast was an increase in the length of the lower masthead from .3 times the topmast length, to .43 times the topmast length, in order to make the lower and upper topsails more equal in size.

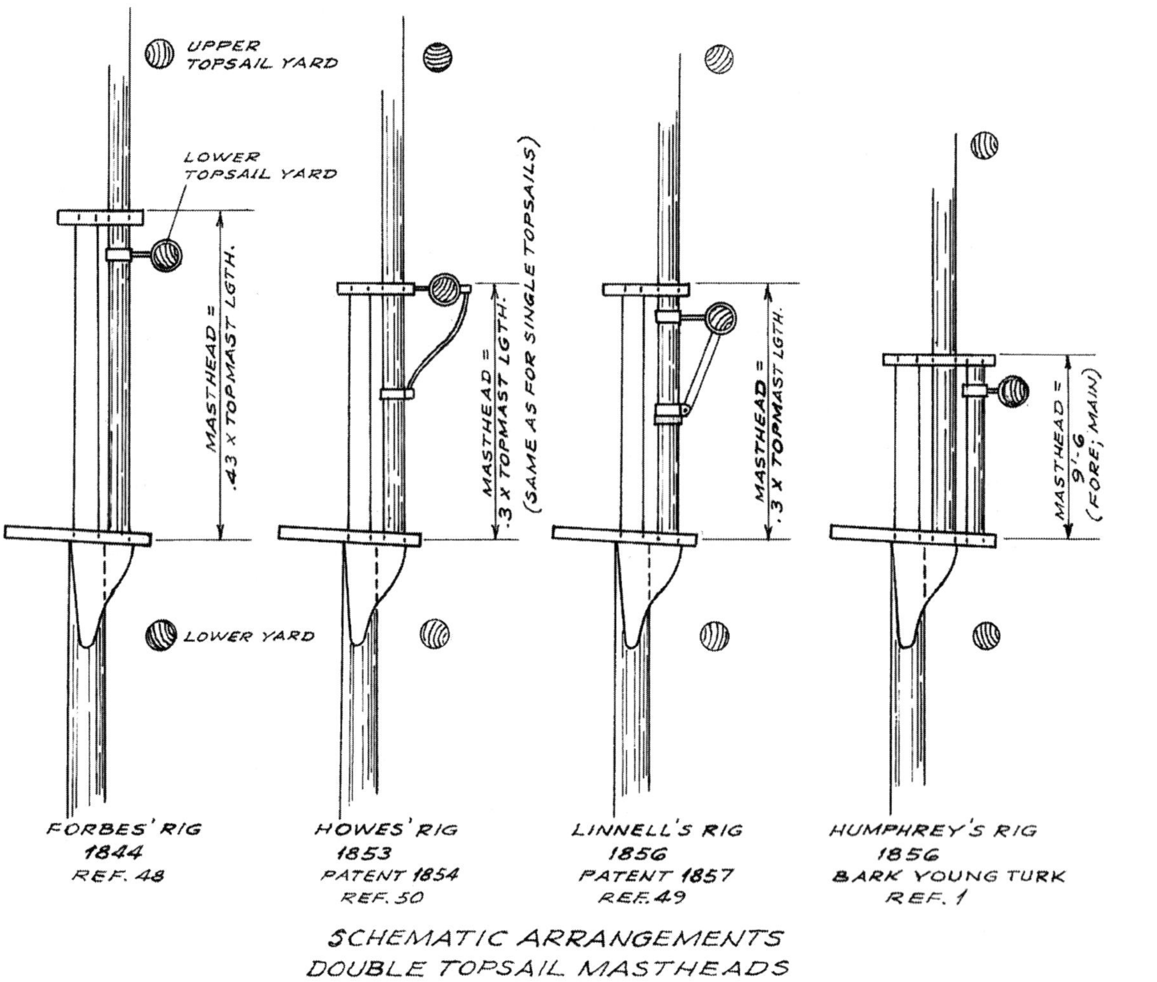

PRINCIPAL DETAILS

FORBES' RIG — THE LOWER MASTHEAD IS LENGTHENED. AN ADDITIONAL HOISTING TOPSAIL YARD IS SECURED BY A PARRAL TO THE TOPMAST BETWEEN THE TOP AND THE LOWER CAP. THE RIGGING IS CONVENTIONAL.

HOWES' RIG — AN ADDITIONAL TOPSAIL YARD IS FIXED TO THE LOWER CAP WITH A SWIVEL. IT IS SUPPORTED BY A CRANE FITTED WITH A PINTLE INTO A FIXED BAND AROUND THE TOPMAST IN THE DOUBLINGS.

LINNELL'S RIG — AN ADDITIONAL FIXED TOPSAIL YARD IS SECURED TO THE TOPMAST BY A LOOSE BAND. THE YARD IS SUPPORTED BY A CRANE REACHING DOWN TO A LOOSE BAND WHICH RIDES ON ANOTHER FIXED BAND AROUND THE TOPMAST.

HUMPHREY'S RIG — A STUB TOPMAST IS STEPPED IN THE TOP FORWARD OF THE TOPMAST. AN ADDITIONAL TOPSAIL YARD IS SECURED TO THIS STUB TOPMAST BY A PARRAL AND FUNCTIONS IN THE CONVENTIONAL MANNER.

Figure 6.9. Double topsail mastheads.

The new lower topsail yard was outfitted in the same manner as the upper topsail yard.

Forbes spent considerable time and effort in promoting his invention but never applied for a patent, being wholly content to see it accepted by the maritime public.

In 1849 he published a pamphlet, "A New Rig for Ships and Other Vessels combining Economy, Safety and Convenience," Ref. 48. In this pamphlet he included the sail plan of a vessel similar to *Samoset*, 1847. This sketch is the only known depiction of a ship with Forbes' invention which was produced in any of his articles and is illustrated in Fig. 6.10.

Opposite, top: **Figure 6.10. A ship with Forbes' Rig.** ***Opposite, bottom:*** **Figure 6.11. Frederick Howes' application for double topsails patent.**

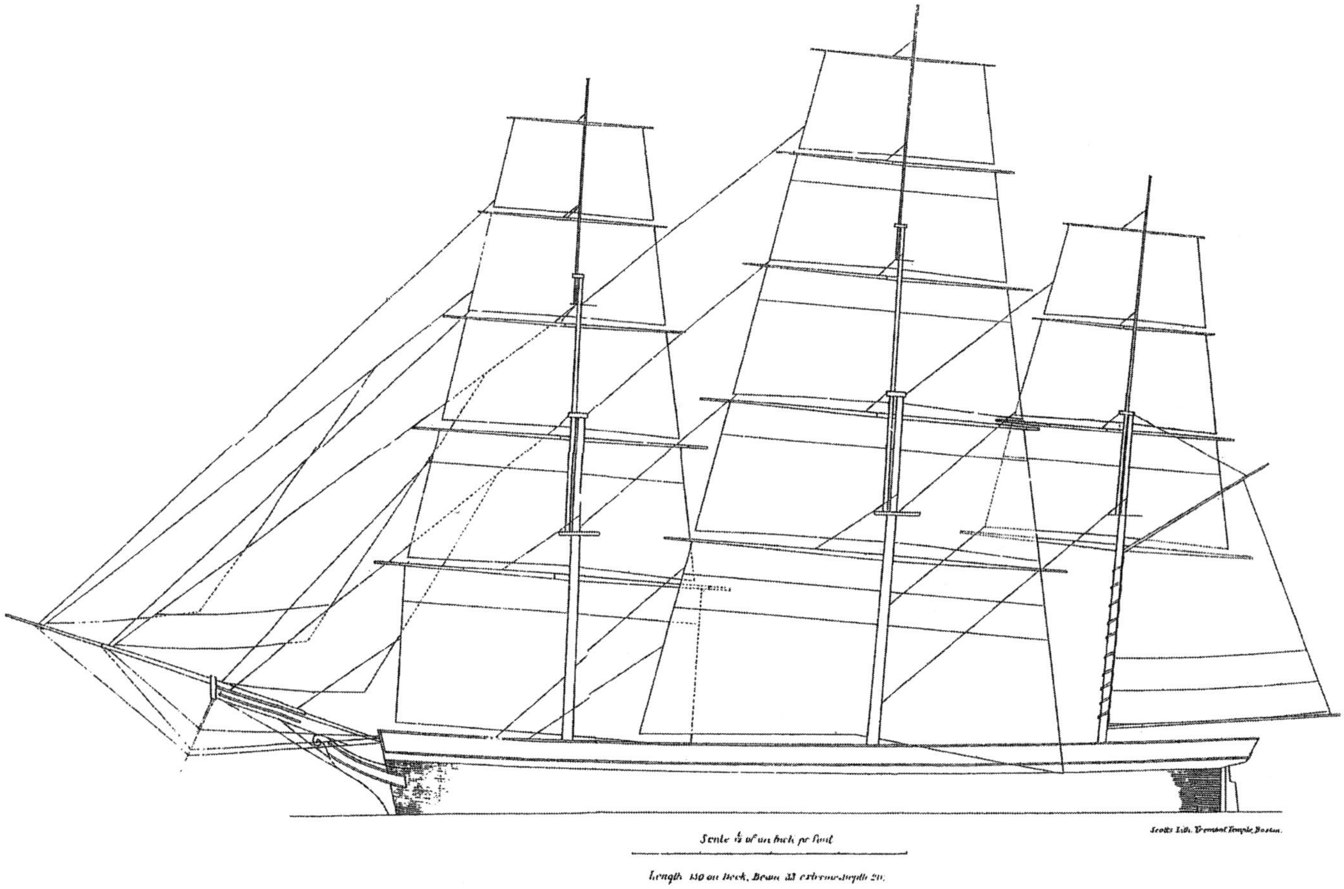

UNITED STATES PATENT OFFICE.

FREDERIC HOWES, OF YARMOUTH, MASSACHUSETTS.

EXTRA YARD TO TOPSAILS.

Specification of Letters Patent No. 11,125, dated June 20, 1854.

To all whom it may concern:

Be it known that I, Frederic Howes, of Yarmouth, in the county of Barnstable, the State of Massachusetts, have invented a new and improved mode of applying and supporting a second or extra topsail-yard by means of a truss to the cap of the lower masthead, and also a crane or brace to the heel of the topmast, top, or trestletrees, to the lower mast; and I do hereby declare that the following is a full and exact description thereof, reference being had to the accompanying drawings and to the letters of reference marked thereon.

The nature of my invention consists in applying an extra yard to any topsail, and supporting it upon the cap of the lower mast head by a truss and crane, as hereinafter described, so that the upper topsail yard may be lowered down in close proximity to the said extra yard. By this improvement the topsail is divided into two parts and may be reefed or taken in with one third the number of men necessary to secure the same result in the old plan. When, as by the old plan, the yards are supported by a chain attached to the topmast above such yard the upper topsail yard cannot be lowered below the point of attachment of such chain to the topmast, and of course the sail will be "bellied out" by the wind and require considerable tackle and a large number of men to take it in, but by my improvements the upper yard may be lowered down to the new yard, and the upper half of the topsail, as divided, drops behind the lower half of the sail, and after it is taken in or furled the lower half is clued in from the deck by the usual tackle and in the ordinary way.

The mode of applying and sustaining my new or extra yard is shown in elevation and in detail in the figures of the accompanying plate of drawings.

To enable others skilled in the art to make and use my invention, I will proceed to describe its construction and operation.

I construct my yard in any of the known forms supported by a truss marked A in the accompanying drawing, attached to the cap marked B, connected to the after part of the yard by iron bands *c, c*, on the forward part by a brace or crane, marked D, the upper end connected to the band marked E, by a swivel head, the lower end of the brace or crane attached to the band marked F, by a pintle.

What I claim as my invention, and desire to secure by Letters Patent, is—

The application of an extra yard, supported by truss, crane, or brace, as herein described, or any other substantially the same, and which will produce the same effect.

FREDERIC HOWES.

Witnesses:

E. E. Rice,
Nathan Edson.

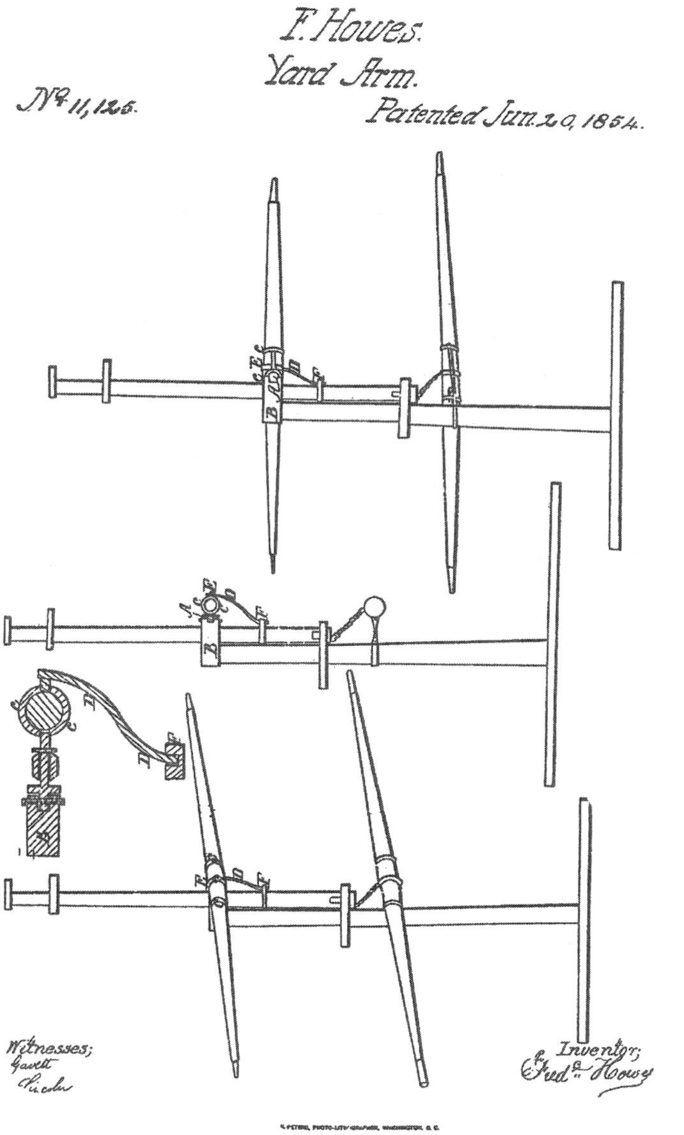

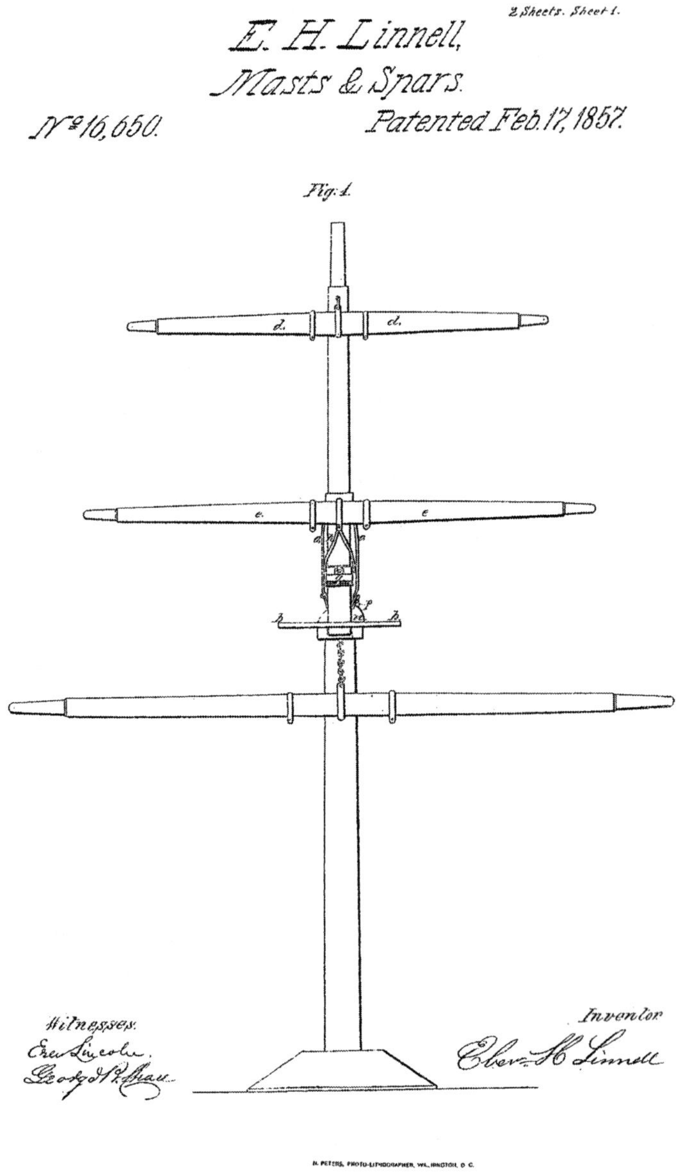

Left: **Figure 6.12. Howes' sketch of double topsail details.** *Right:* **Figure 6.14. Linnell's Arrangement of double topsail yards.**

In 1853, Frederic Howes, of Yarmouth, Massachusetts, introduced a new version of double topsails in the clipper *Climax*. In this case the lower masthead was not modified. The lower topsail yard was trussed directly onto the forward end of the cap by means of a swivel head which attached to the after side of the yard by an iron band.

The forward side of the yard was supported by an iron brace which reached down to an iron band that was fitted around the topmast and was made to receive a pintle on the foot of the brace.

This yard could not be raised or lowered but was otherwise fitted with conventional topsail yard rigging. It is illustrated in Fig. 6.9.

Howes, unlike Forbes, was interested in protecting his invention and so applied for a patent which was granted on June 20, 1854. A copy of his application and an accompanying sketch are included herein. See Figs.6.11 and 6.12.

Two additional versions of the double topsail came onto the American scene, both in 1856. Eber H. Linnell, of Orleans, Massachusetts, fitted the ship *Flying Mist* with a lower topsail yard that was attached to the topmast; and, in Boston, John Humphrey outfitted the bark *Young Turk* with an additional stub-topmast which was stepped in the lower tops and to which an additional topsail yard was slung.

In the case of Linnell, the lower topsail yard is supported by two loose collars which are free to rotate on the topmast. The upper collar is secured to the yard and is joined to the lower by iron braces. The lower collar rests on a permanent

Opposite: **Figure 6.13. Eber Linnell's application for double topsails patent.**

UNITED STATES PATENT OFFICE.

EBER H. LINNELL, OF ORLEANS, MASSACHUSETTS.

ATTACHING EXTRA TOPSAIL-YARDS TO VESSELS.

Specification of Letters Patent No. 16,650, dated February 17, 1857.

To all whom it may concern:

Be it known that I, E. H. LINNELL, of Orleans, in the county of Barnstable and State of Massachusetts, have invented certain new and useful Improvements in Rigs for Vessels, and that the following description, taken in connection with the accompanying drawings, hereinafter referred to, forms a full and exact specification of the same, wherein I have set forth the nature and principles of my said improvements by which my invention may be distinguished from others of a similar class, together with such parts as I claim and desire to have secured to me by Letters Patent.

The figures of the accompanying plate of drawings represent my improvements.

Figure 1 is a front elevation of a mast, &c., with my improvements applied thereto. Fig. 2 is a central vertical section of the same.

The present invention consists in a new mode of attaching and applying extra top-sail yards in square rigged vessels. Extra top sail yards have been sometime in use, and have heretofore been supported by a chain, and by more recent improvements, some of which are now very generally applied, by a swivel joint in a ring attached to the cap, with a vertical arm resting on the trestletree. Among the disadvantages of suspending the yard from the cap in this way are the following:—The whole strain comes upon the cap, which ordinarily, without this addition, sustains all the strain it ought to bear, while the yard itself, in order to be attached snugly to the cap, has to be made smaller than is consistent with strength, so that in a gale of wind the yard is often broken. The head rope also of the lower top-sail, swings round unequally with the foot rope of the same, as the extra top-sail yard being attached to the cap, swings from a center that is in a different plane from the lower yard, thereby twisting and wrinkling the sail and straining the clews. Again by this mode of attachment to the cap, the yard cannot be braced up sharp because it interferes with and chafes the top-mast rigging. By my improvements all these objections are entirely obviated the lower top-sail yard being suspended and swinging from the top-mast as a center, thereby bringing all the yards in the same diagonal plane and preventing the sail from wrinkling, as the head rope of the top-sail swings around equally with the foot rope of the same. By attaching the yard to the top-mast itself, the cap is relieved from all strain, and the yard instead of being made smaller as is necessarily the case when attached to the cap, can be made of any desired size and strength. By my mode of attachment also the yard can be braced up sharper than by other rigs, as when so braced, it hangs entirely clear of the top-mast rigging and cannot chafe the same. The advantage of having the yards all in the same plane, so that the sails will all draw smooth and free from wrinkles will readily be manifest.

a a in the drawings represent the top-mast, *b b* in the trestletree and *c* the cap.

d d is the upper top-sail yard and *e e* the lower top-sail yard. This yard *e e* is attached to the top-mast *a a* by collars *f* and *g* the upper collar *f* being attached to the yard and the lower one to the diagonal forked brace *h* which sustains the yard in position. Both the collars *f* and *g* encircle and turn on the top-mast, the lower one *g* turning upon an annular plate or bearing *h′* set into the periphery of the top-mast a little. The yard *e e* is attached to the upper collar *f* by a swivel joint *i* and to the forked brace *h* by a similar joint *k*. The upper yard *d d* is attached to the top-mast by a collar *l* and swivel joint *m*.

From the foregoing description it will be seen that the yard swings from the top-mast as a center and that the yard is supported entirely by the top-mast thereby relieving the cap of all strain, and preventing the head of the mast from being twisted. By this mode of attachment as the yard stands out from the top-mast, instead of being drawn in close to the same, as is necessarily the case when suspended from the cap, sufficient room is afforded for the yards to be braced sharp up without interfering with the top-mast rigging, and this mode of attachment also causes the lower top-sails when the upper yard is hoisted to hang in the same plane from the upper top-sail yard or the foot of the top-sail.

In order to relieve the strain upon the fid *n*, occasioned by the additional weight of the extra yard and its attachments, diagonal braces *o*, *o*, one on each side are attached to the mast head, and fastened to a

band *p* passing through the foot of the top-mast, thereby partially sustaining the top-mast, and relieving its weight upon the fid.

The forked brace *h* performs an essential service in securing the easy management of the yard, as the two forks of the brace attached to the turning collar balance the yard, while the collar permits the free play of the yard and brace.

Having thus described my improvements I shall state my claim as follows:

I do not claim the use of an extra top-rail yard, nor the holding it stationary, *i. e.*, so as not to be raised or lowered, as both the use of an extra top-sail yard, and so holding it have long been known, and practised, but

What I do claim as my invention and desire to have secured to me by Letters Patent is—

Attaching and hanging the extra top-sail yard to the top mast itself, by means of the combination of the collars *f* and *g* turning on the top mast, the forked brace *h* connecting said collars and the stationary annular plate *n′* on which the lower collar *g* rests and turns, as herein above described, so that said extra toprail and yard shall swing from the axis of the top-mast as a center, and secure the advantages herein above specified.

EBER H. LINNELL.

Witnesses:
EZRA LINCOLN,
GEORGE R. CHASE.

fixed band attached to the topmast.The entire assembly rotates when the yard is braced.

With both Linnell and Howes the yard functions at a fixed height.

Linnell applied for and was granted a patent on February 17, 1857. A copy of this patent and its accompanying sketches are included herein as Figs. 6.13, 6.14 and 6.15. A schematic arrangement of Linnell's invention appears in Fig. 6.9.

The final venture into proposed double topsails was undertaken by John Humphrey, in Boston, in the small, 350 ton, bark *Young Turk*, in November 1856. This consisted of stepping a short stub-topmast in the forward end of the top, forward of the traditional topmast, and terminating in the cap. The new, additional yard, was then slung from this stub-topmast.

A reconstruction of this invention is shown in Fig. 6.9 and is based on a description of the vessel as given in Ref. 1 and which is repeated below.

> She has a new rig invented by Capt. John Humphrey. It consists of double topmasts and double topsail yards. The lower topsail has a single reef in it and the lower topmast is just long enough to set this sail. The second topmast has crosstrees, doublings and cap, and is fitted for the upper topsail which also contains a single reef. When the upper topsails are furled, the vessel is then under the same surface of canvass [sic] represented by double reefs in the old rig, and when the lower topsails are reefed, she is under close reefed topsails.

This, apparently, is the only known reference to this version of double topsails.

The Howes version of double topsails was accepted and adopted as the standard. Ironically, however, it is Forbes' name that is almost universally associated with the double topsail rig.

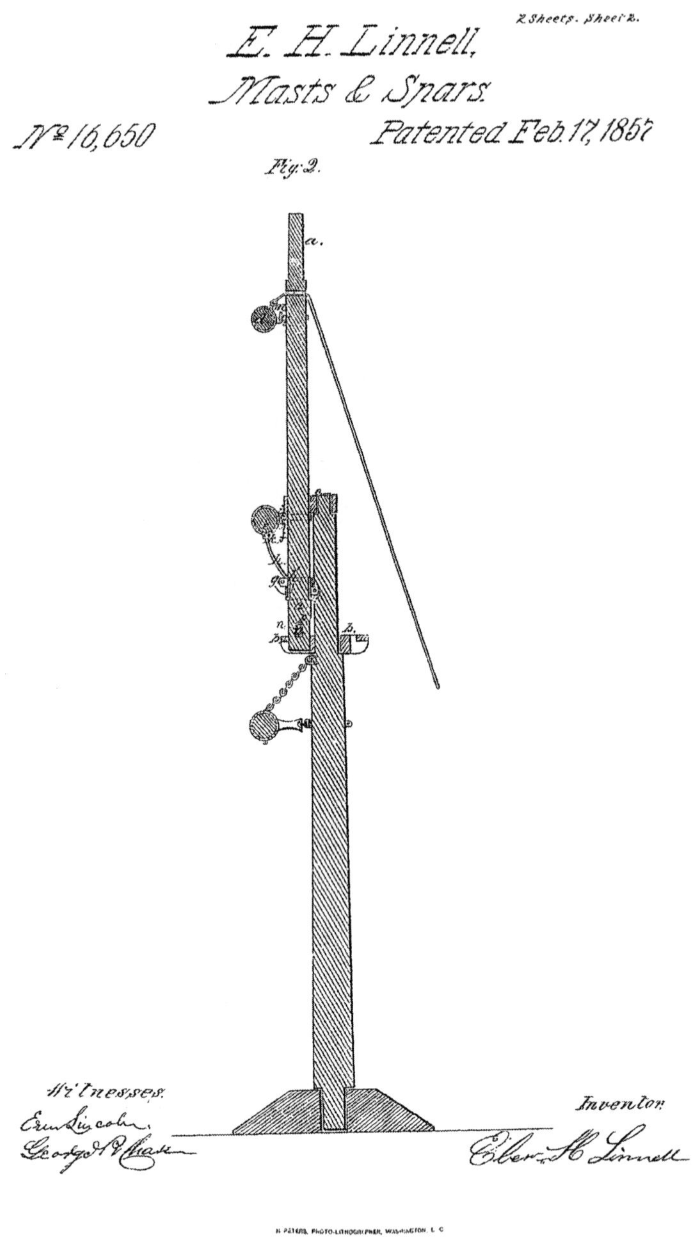

Figure 6.15. Details of Linnell's double topsail yards.

7

Upper Masts

The previous chapter dealt with making the lower masts and bowsprits, the last remaining large structures to be fitted in a ship. The various spars still to be accounted for all contributed toward the viability of the vessel, each in a proportionate way. In each case the scantlings of the items were of no great magnitude but, separately, each one was a vital part of the whole.

In order to provide a ship with motive power additional masting was required and, by the nature of a ship's form and structure, masts could only proceed upward. At first glance this choice would seem to be without limit but there were adverse consequences accompanying this selection.

This consideration is clearly covered in the following statement by Kipping.[33]

> The exact height of the masts in proportion to the form and size of the ship remains yet a problem to be determined. The more the masts are elevated above the center of gravity the greater will be the surface of the sail which they are enabled to present to the wind,—so far an additional height seems to be advantageous. But this advantage is diminished by the vibrating movement of the mast, which operates to make the vessel stoop to its effort; and this inclination is increased in proportion to the additional height of the masts—an inconvenience which it is necessary to guard against; thus what is gained upon one hand is lost upon the other. To reconcile these differences, it is certain that the height of the mast ought to be determined by the inclination of the vessel, and that the point of her greatest inclination should be the term of the height above the center of gravity.

Consideration of these consequences resulted in the topsail becoming the largest sail in a ship and the succeeding sails above were a progression of diminishing sizes. The tauntness of the masts and the squareness of the yards resulted in diminished weights aloft.

The proportionate lengths of masts and yards are recorded in Table 4.1.

All masts above the lower masts were intended to be operable in the sense that they were fidded in a manner which allowed them to be struck (lowered) when the situation warranted.

The diameter of topmasts was generally about half the diameter of their lower mast. Other upper masts were proportionately smaller. See Table 4.1.

To Make a Topmast

The length of a topmast is given as the hounded length plus the headed length or, as the whole length including the head, which also includes the tenon for the cap. The thickness of these parts must be able to pass through the round hole in the lower cap.

The size of the topmast, both diameter and length, are generally taken from tables compiled for ships of a given size.

A stick of appropriate dimensions is examined and, if found suitable, made straight and four-square for its entire length. The butt end of the stick is the heel; the upper end is the head.

The middle line of the stick is struck along one square for its entire length. Across this line, starting from the heel, are struck lines at the location of the lower cap, the locations of the four quarters between the lower cap and the hounds, and the head, exclusive of the tenon. (The tenon has, however, been included in the total length.)

The given diameter of the mast is at the location of the lower cap. This diameter is maintained between the cap and the heel of the mast. See Fig. 7.1, 1st trim.

Starting from the cap the value of each quarter between the cap and the hounds is marked on each appropriate quarter. These values can be taken directly from Table 4.6 or may be calculated by using the fractional proportions which appear in the table. The sizes of the topmast head are taken off in similar fashion from this table.

A fair curve is struck through the points thus gathered, a curve on each side of the middle line, between the cap and the hounds. The configuration thus obtained is the shape of the topmast.

The four sides of the stick are converted to this configuration using the same method that was used in converting the single tree mast in Chapter 5 and shown in Fig. 5.1.

Next, two stops (shoulders) are marked on the stick. One is placed two and one-half diameters of the mast above the heel; the other is located at the top of the hounds. These are shown in Fig. 7.1, 3rd trim. They retain the full diameter of the stick which is now converted to eight-square from the lower stop to the upper stop. The head remains square and the heeling below the lower stop is chamfered on the four corners. If the heeling does not fill the hole between the trestletrees, fillings or shims are fayed and nailed to the heel to make up the deficiency, allowing one-quarter inch of play.

The mast is now rounded between the lower stop and the lower end of the hounds which remain eight-square.

Next, a squared hole is cut athwartships for the fid, its lower edge to be the depth of the trestletrees above the heel plus one inch for an iron plate that is fitted on the trestletrees. The hole is sized to accept an iron fid which is, in depth, one-third the diameter of the topmast, and in width two-thirds the depth of the fid. The fid, in length, is one and one-half diameters of the lower mast. See detail in Fig. 7.1.

Immediately above the lower stop a sheave-hole for the top-rope is cut transversely through the topmast. It enters in the foremost eight-square on the starboard side and exits on the aftermost eight-square on the port side.

The top-rope is used to hoist the topmast into position parallel with the lower mast.

Cutting the sheave-hole as described applies to the foremast. For the mainmast the hole is cut from the aft starboard eight-square and exits on the forward port eight-square. This arrangement facilitates getting both topmasts up simultaneously.[39]

A hole is bored fore-and-aft through the middle of the hounds for the topsail tye.

The final trim is to chamfer the corners of the topmast head in way of the eyes of the rigging.

Fig. 7.1 illustrates the steps in making a topmast.

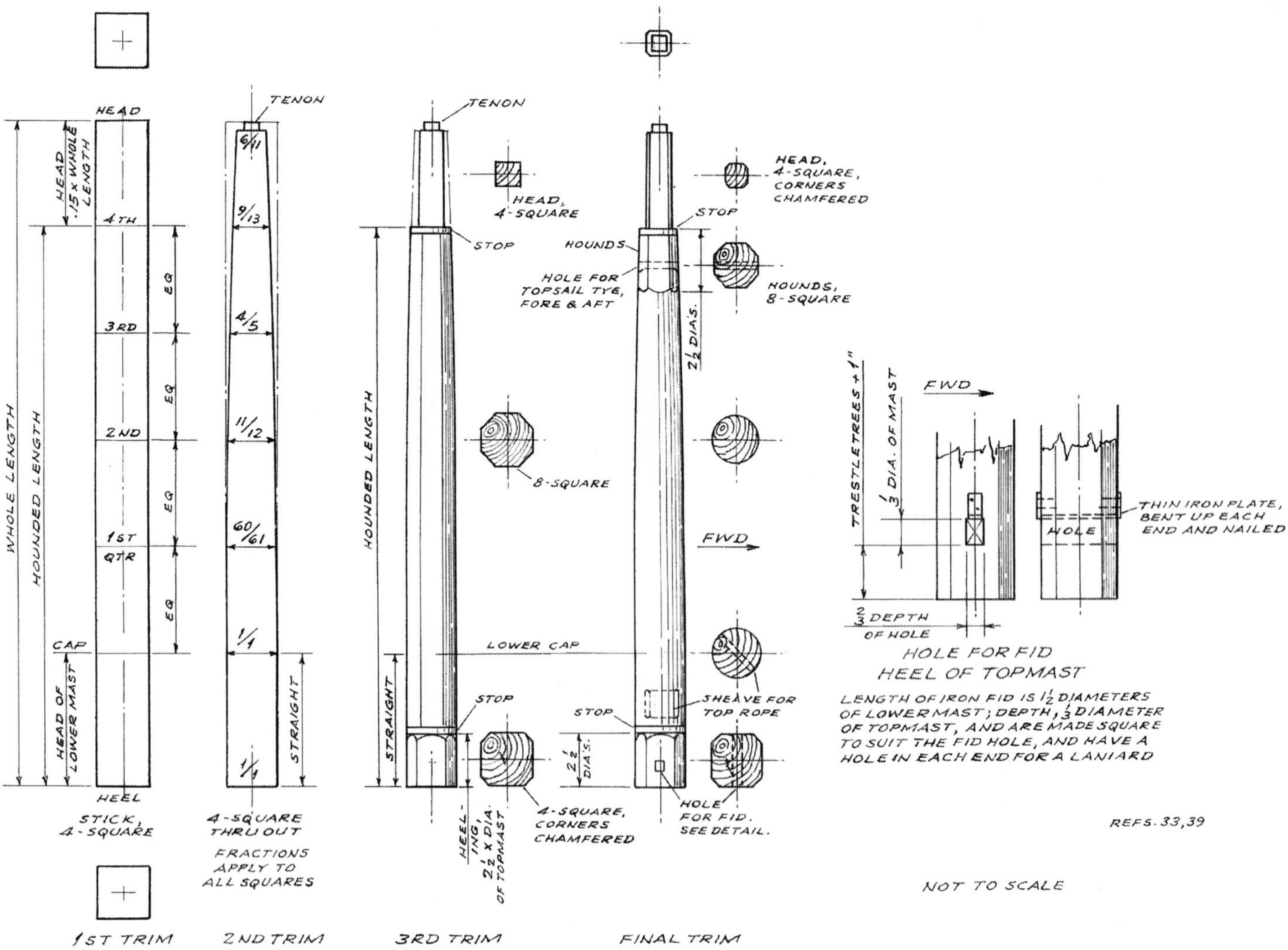

Figure 7.1. To make a topmast.

Furniture for a Topmast

The topmast required the installation of crosstrees and the provision of a cap, to be complete. These items are illustrated in Fig. 7.2 and are described as follows.

The trestletrees (a) are installed one on each side of the masthead and are settled down on the top of the hounds, their inclination fore-and-aft making due allowance for the rake of the masts. The breadth of trestletrees is 7/12 of the breadth of the lower trestletrees of the respective masts, and the depth is 7/12 of the depth of the lower trestletrees. The forward end of each is rounded up and the after end is snaped on the underside. Their length is determined by inclusion of a short crosstree (c) at the forward end; then a chock (g) which fays against this crosstree; then allowance for the heel of the topgallant mast plus one-quarter inch play; next the forward crosstree (b); then the topmast, fitted snugly between the crosstrees and the trestletrees; another short crosstree (c) which restrains the topmast on the after side; a space to allow a block as a fairleader for a topgallant stay; the after crosstree (d); and an overhang of about the width of the crosstree. They are middled fore-and-aft about the centerline of the topmast.

The forward crosstree (b) and after crosstree (d) are the breadth of the trestletrees and, in thickness, 4/5 of their breadth. The length of the after horn (d) is 4/6 of the lower after crosstree of the respective masts; the length of the forward horn (b) is 5/6 of the after one. The horns sweep aft about nine inches in 16 feet. The crosstrees are tapered for about one-third their length on the underside of each end. The ends are rounded. About three inches in from each end a hole is bored for the topgallant shrouds. They are both placed over the trestletrees as shown in Fig. 7.2. A score is taken out of each member to bring the crosstrees down almost flush.

A short crosstree (c) is let into the trestletrees immediately aft of the topmast head to confine the topmast. Another short crosstree (c) spans the trestletrees at the extreme forward end to bind them together and stabilize the entire assembly.

A small chock (g) is fitted aft of the foremost short crosstree. It is of a size to confine the topgallant mast in its proper alignment.

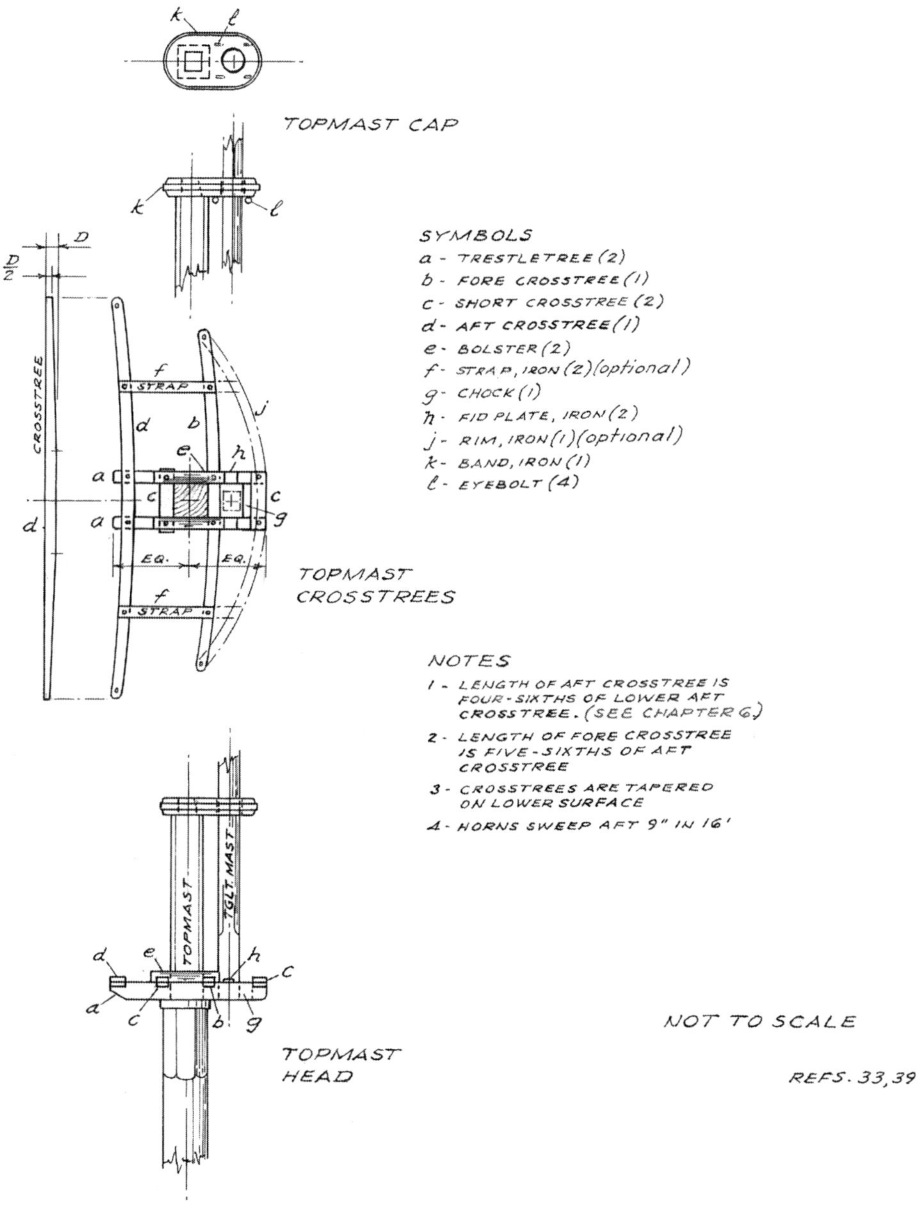

Figure 7.2. Furniture for a topmast.

Two bolsters (e), fir, are let over the trestletrees in way of the topmast head. They are quarter-round and of width one inch greater than the trestletrees. Their length is from the fore crosstree (b) to the aft short crosstree (c), extending a short distance beyond each of these items. They are scored at each end to allow the bolster to fay with the upper side of the trestletrees and are secured with a nail each end into the trestletrees. Their purpose is to ease chafing of the eyes of the shrouds around the topmast head.

An iron strap (f) is sometimes fitted between the fore and after crosstrees to stiffen the assembly, one on each side of the topmast. These are placed outboard about two-thirds the length of the forward crosstree (b). When a semi-circular rim is part of the assembly these straps are extended to include the rim.

Fid plates (h) are small iron plates nailed on the trestletrees in way of the heel of the topgallant mast to protect the trestletrees from wear by the fid.

A semi-circular rim (j), generally of iron, was sometimes installed across the forward end of the crosstrees. It was secured to the forward end of the trestletrees and engaged each end of the forward horned crosstrees. This item was made up of three pieces, the piece in the middle uniting the ends into a single unit.

The topmast cap, made of elm, is the same in proportion to the topgallant mast as the lower cap is to the topmast. The round and square holes are formed in the same manner as in the lower cap.

An iron band (k) encircles the cap to reinforce it. In width the band is one-third the thickness of the cap, and in thickness it ranges between one-quarter and one-half inch depending upon the overall size of the cap.

Eyebolts (1) were driven into the cap from the underside. There are four, two of which are placed on the forward side of the round hole and two are placed correspondingly aft of the round hole. They are driven with the eyes fore-and-aft.

To Make a Topgallant Mast with Pole Masts Above

At the period of this book it had become usual practice to combine the topgallant, royal and skysail masts into a single continuous pole, as opposed to a royal mast which was fidded to the topgallant mast.

In this pole arrangement the topgallant and each succeeding mast above were built with a stop at its head rather than the conventional treatment of hounds, crosstrees and squared head to provide a doubling. The stop was merely a shoulder formed by the lesser diameter of the higher mast. This shoulder prevented the eyes of the rigging and the collar of the stay from sliding down along the mast.

The topgallant mast, like the topmast and lower masts, was made in the same manner as these latter two. First, a tree of economical size was converted into a stick four-square and of length to accommodate all the divisions of the assembly from the heel of the topgallant mast to the truck. This could result in a slender pole 75 feet long.

Starting from the topgallant heel a middle line was struck along the length of the stick. Then, also from the heel, the length of the topmast doubling was lined across the stick. This point, the location of the topmast cap, was the extent of the given diameter of the topgallant mast.

Next, the distance between this point and the topgallant stop was divided into four quarters. A line was struck across the stick at each quarter. From tables, such as Table 4.6, the numerical value of each quarter was marked and a line was struck through all the marks on each side of the middle line. These lines depicted the shape of the mast from heel to stop. This is shown in Fig. 7.3.

The stick was now canted until the markings were repeated on two sides. This allowed the stick to be shaped four-square by using plumb lines in the manner illustrated in Fig. 5.1.

The stick was next converted to eight-

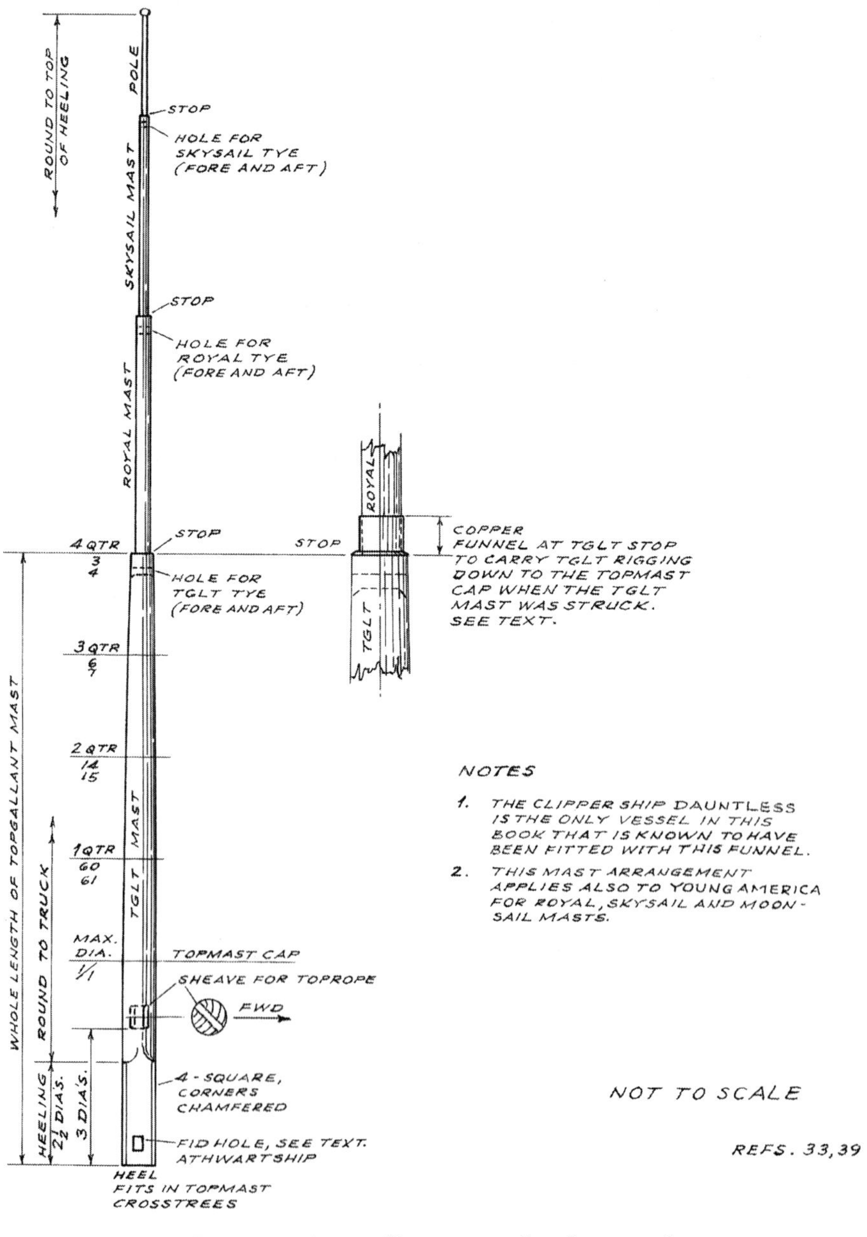

Figure 7.3. A topgallant mast with pole masts above.

square throughout its topgallant length except for a two and one-half diameter length at the heel. The final trim converted the eight-square to round, and the square heeling was chamfered on the four corners.

On some occasions the topgallant mast, from the stop, down for a distance of two diameters, was gently faired out as an eight-square until, at the stop, it was the same dimension as the second quarter of the mast. This provided greater landing for the rigging on the stop. This is illustrated in Fig. 7.4.

Above the topgallant mast the royal and skysail masts were marked at their stops and then developed in the same manner as the topgallant mast. The royal mast, even with its small diameter, was quartered and shaped to dimensions which appear in Table 4.6.

Above the royal stop the skysail mast was made round for its entire length. It was reduced in diameter at the foot in order to provide a stop for the royal rigging. At the head it was probably further reduced and, between these two extremes, it was round throughout and diminished in a straight taper.

The tables of sizes and proportions of masts that were used in the 1850s were compiled many years previously and contained values up to and including the royal mast only.

Above the skysail mast a pole provided· a stop for the skysail rigging. Poles came in three variations: the long pole, an extremely long, slender pole; the stump pole, a short, stubby stick; and the common pole, which was of a middle length. All provided a necessary stop for the skysail rigging and all were fitted with a flag halliard. This arrangement is shown in Fig. 7.3.

Several holes were bored through this finished mast assembly. At the heel of the topgallant mast a fid hole was bored athwartships through the mast. The bottom edge of the hole was one diameter plus one inch up from the heel; the vertical dimension was one-half the mast diameter; and athwartships it was two-thirds of that size. The fid was one and one-half times the diameter of the topgallant mast in

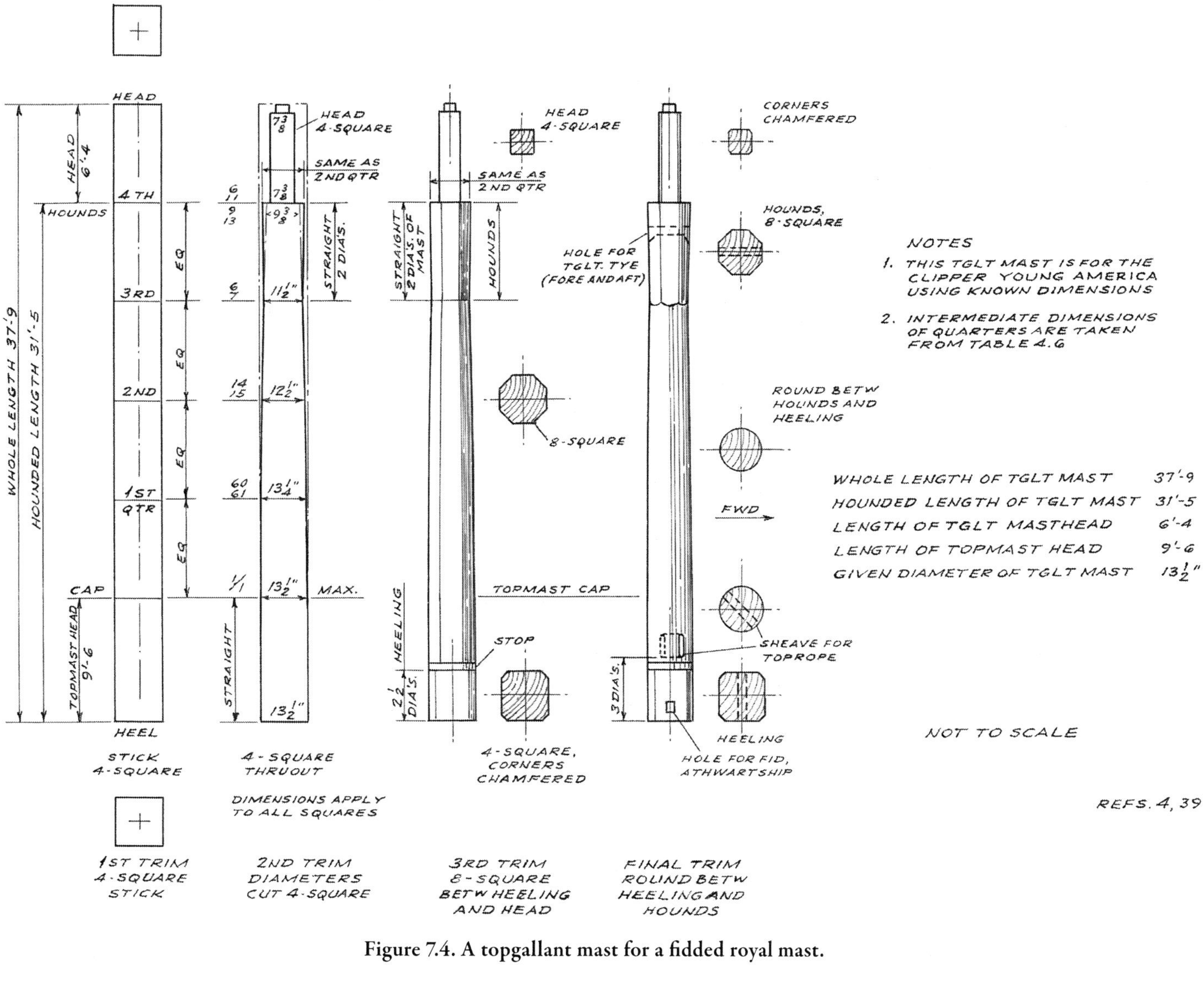

Figure 7.4. A topgallant mast for a fidded royal mast.

length. Its cross-section was of size to fill the fid hole. In each end was a hole for a laniard.

Next above the heel was a sheave hole for the top rope. This hole, one diameter and one-sixth in length, entered the mast from the foremost eight-square on the starboard side and exited on the aftermost eight-square on the port side. The lower end of this hole was three diameters above the lower end of the mast.

Finally, there were sheave holes cut in a fore-and-aft direction, one half the diameter below each stop. These were, in length, one diameter plus one-sixth, and were lined with copper.[33] These holes were each for the appropriate yard tyes.

A feature rarely found in the merchant service was the installation of a funnel over the topgallant stop, man-of-war fashion.[1] This was a copper sleeve fitted around the lower end of the royal mast with a sliding fit. Its bottom edge was flanged all around to rest on the topgallant stop. It was long enough to engage all the eyes of the topgallant rigging, usually three each, port and starboard. Its purpose was to retain the topgallant rigging on the topmast cap when the topgallant mast was struck. It is shown in Fig. 7.3.

The clipper ship *Dauntless* is the only ship included in this book which is known to have had this funnel installation.

To Make a Topgallant Mast with a Fidded Royal Mast Above

In the 1850s the topgallant, royal and skysail masts were combined into one single pole graduated with stops, as in Fig. 7.3. The earlier system of making the topgallant mast a separate stick, complete with masthead and crosstrees, had been discarded because of its weight and expense. However, it had not disappeared entirely.

The clipper ship *Young America*, which is included in this book, was designed to carry moonsails above her skysails. This would have resulted in a pole 85 feet long with a maximum diameter at the heel of the topgallant mast of only 13½ inches. A stick this slender did not seem to be a wise decision so her builder, William H. Webb, opted to shorten it by making the topgallant mast a separate stick complete with a masthead and crosstrees. The resulting mast is detailed in Fig. 7.4. The dimensions are taken from her masting table which is included in the Appendix.

This mast, like all others, starts as a stick, four-square

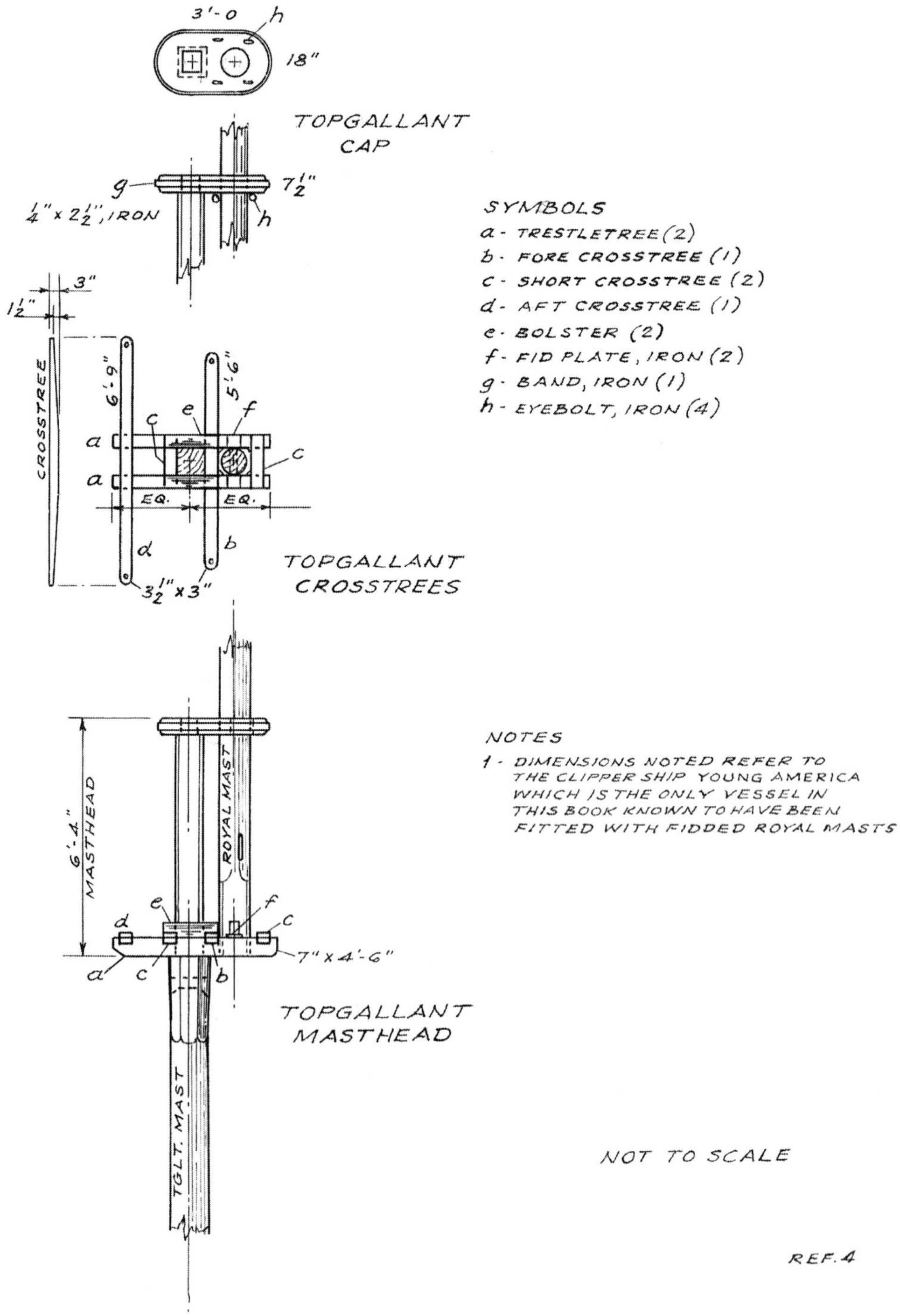

Figure 7.5. Furniture for a topgallant mast with fidded royal mast.

and out of winding. The quarters and a middle line are lined on the surfaces as required and the proportional diameters taken from Table 4.6. See Fig. 7.4, 2nd trim for details.

Before the stick is cut to the diameters noted the upper end, in way of the hounds, is modified as follows. The masthead is marked in above the hounds. From the foot of the masthead, which is the stop, down a distance of two diameters of the mast a line is struck on all four surfaces. Also at the foot of the masthead the diameter of the second quarter is marked. Between these two markings the hounds are gently flared out in an eight-square. This provides an increase in the surface of the stop. The remainder of the mast, between the heeling and the lower of the above lines is converted to eight-square as in the third trim. The masthead is also trimmed to size.

In the final trim the eight-square, from the heeling to the lower end of the hounds, is converted to round and the heeling and head are chamfered on the corners.

A fid hole is cut athwartships through the heeling. The lower edge of the hole is one diameter plus one inch above the heel. The hole, up and down, is half the mast diameter; and athwartships is two-thirds of the up and down.

Next above the fid hole is the hole for the top rope. This is located in the starboard foremost eight-square and exits in the port aftermost eight-square. Its lower edge is three mast diameters above the lower end and its length is one diameter and one-sixth.

Finally, there is a hole for the topsail tye, cut fore-and-aft, half the diameter below the masthead.

Furniture for a Topgallant Mast with a Fidded Royal Mast

The furniture required for the headed topgallant mast is essentially the same as that fitted to a topmast except that its scantlings are smaller, in keeping with the smaller dimensions of the mast being outfitted. See Fig. 7.5.

The sample used here is reconstruction based on known dimensions of the clipper ship *Young America*.

Omitted from this crosstree arrangement are the fore-and-aft iron straps that connect the aft crosstree of the topmast with the forward one; also omitted is the curved forward rim of the mast assembly which may have been a routine option.

8

Yards

Yards occur in two variations, square and lateen; the first are suspended across the mast at right angles and the second cross the mast obliquely. The square yards are the variation that occur in this book and were in general commercial use during the period.

They are of cylindrical form tapering from the middle, which is termed the slings, toward the extremities which are termed the yardarms. The slings are the location of the given diameter of the yard. The distance between the slings and the outer end (the yardarm) is divided into four equal parts, called quarters. Starting from the slings they are identified as the first, second and third quarters, and the yardarms.

Lower yards are the largest, both in length and diameter and, in most cases, can be made from a single tree. However, in the largest vessels, it was probable that the fore and main lower yards would be made from two trees scarphed at the middle with a vertical scarph. After the scarph was completed it was reinforced by iron hoops which encircled the yard at intervals of about two feet six inches, with the end hoop being placed about nine inches within the nibs of the scarph. One hoop was placed in the middle of the scarph for attachment of the chain slings which lead over the masthead.

It was also customary to place a hoop over any large knot to reinforce the yard at that particular location. This applied to single tree yards as well as those that were scarphed. Also, the scarphs were coaked.

The taper of the yards, from slings to yardarms, was standardized in the same manner as those that applied to lower masts. The values of these proportions are shown in Table 4.8.

The steps required to make a single tree yard and a scarphed yard are illustrated in Figures 8.1 and 8.2, and the fittings required for these yards are shown in Figures 8.6 and 8.7.

The following general information applies to lower yards. Length of main yard = 7/8 of mainmast; dia. = 7/10 inch to three feet. Length of fore yard = 7/8 of main yard.

Proportions of diameters to lengths = one inch for every three feet. Proportions of small diameters to given diameters = 1/2 at the yardarm.

Diameters of yards at the quarters—Slings = 1

1st quarter = 30/31
2nd quarter = 7/8
3rd quarter = 7/10
Yardarm = 3/7

In general practice not all of these rules were followed religiously. In most cases a builder's personal philosophy prevailed.

The actual operations that were employed in converting a four-square tree into a finished yard were the same as those used to convert a four-square tree into a lower mast. These operations are described in Chapter 5 and illustrated in Fig. 5.1.

To Make a Lower Yard from a Single Tree

There is very little information available that contains details of the construction of the lower yards which were fitted in specific vessels. The masting tables that do exist generally give the sizes of yards, i.e., length, diameter and yardarm, but no additional details. One can only surmise as to whether such information dictates the use of a single tree for a yard or a scarphed yard made of two trees.

The clipper ship *Great Republic* had lower yards that were 26 and 28 inches in diameter, in which case it seems reasonable to assume that these yards were constructed of multiple trees and were special unto themselves. Unfortunately this assumption cannot be verified.

This, however, was a unique instance, and in the general run of large vessels 24 inches was a large diameter for a lower yard.

Yards, like masts, start as trees which are converted to four-square sticks and go through various stages of trim. The procedure for a yard made from a single tree was as follows and is shown in Fig. 8.1.

First trim—The selected tree was made four-square and out of winding to the desired diameter and length after which a middle line was struck along the length of the stick. At

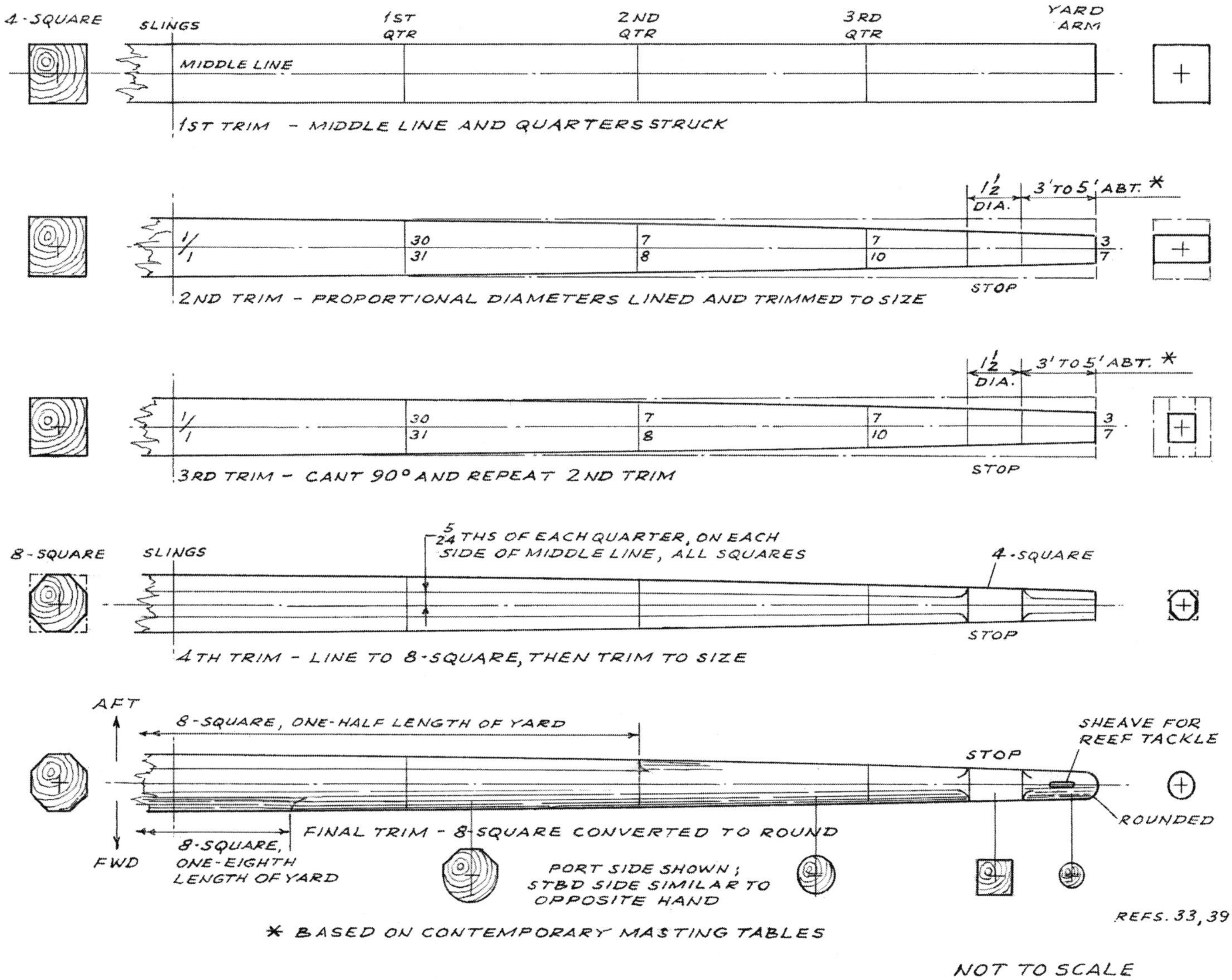

Figure 8.1. A lower yard made from a single tree.

mid-length of the stick a line was struck across the surface from one side to the other. This was the location of the slings and was the maximum (given) diameter of the yard. Working in each direction from the slings the yard was divided into quarters which were lined square across the stick from one side to the other.

Second trim—Proportional sizes were then marked on the quarters and yardarms and connected to show the configuration of the yard. The yard was trimmed to these lines. The stop was next marked outside of the third quarter.

Third trim—The stick was canted 90 degrees, marked as in the second trim, and trimmed to size. This resulted in a four-square stick, the sections corresponding to the proportional values at any point.

Fourth trim—New middle lines, quarters and yardarms, and the stop were struck on the exposed surfaces, all sides. An eight-square configuration was lined on each surface by measuring 5/24ths from each side of the middle line at each quarter and the yardarm. These points were then connected on all sides. The stop remained four-square. The stick was then planed to these lines which resulted in an octagonal shape throughout the length of the yard, except for the stop.

Final trim—The eight-square was converted to round as shown in Fig. 8.1, Final trim.

On the aft side of the yard the eight-square was retained for one-half the length of the yard; the forward side of the yard remained eight-square for one-eighth the length of the yard. The stop remained four-square.

A lower yard made from a single tree is shown drawn to scale in Fig. 8.6.

To Make a Lower Yard from Two Trees, Scarphed

When a yard could not be economically made from a single tree, or when a tree of sufficient size was not available, it was customary to unite two trees of smaller size into a single unit by means of a plain nib scarph.

In actual practice there was probably great variation among shipbuilders as to what details should cover the uniting of the two selected trees. However, in theory, certain proportions were prescribed for the scarph and these are employed in the following description[33,39] which is augmented by Fig. 8.2.

First trim—Two trees of selected diameter were made four-square and out of winding. The length of each tree was five quarters plus four feet, as shown in Fig. 8.2, first trim. Note that the slings are located one quarter plus four feet from the butt end of each selected tree. The top of the tree formed the yardarm. The locations of the slings, quarters and yardarm were lined across the top surface.

Second trim—Two trees. A middle line was struck. The scarph was then laid off in way of the slings. From the slings the first quarter plus four feet on each side was the length of the scarph which was nibbed three inches at each end, as shown in Fig. 8.2, second trim. The stops were next laid off. These were located three feet to five feet from the end of the yardarm.

Third trim—Two trees. The scarph was cut in each tree.

Fourth trim—Combine the trees into one stick, four-square throughout its entire length.

Fifth trim—One unit. Strike proportional dimensions at the quarters and yardarms then trim to these dimensions. All sections are now rectangular except the slings, which remain square, as shown in Fig. 8.2, fifth trim.

Sixth trim—The tree was then canted 90 degrees so that the scarph was horizontal. The fifth trim was repeated making the stick four-square throughout.

Seventh trim—Line all sides to eight-square by measuring 5⁄24 from the middle line on the slings, each quarter and the yardarms, except the stops which remain four-square. Cut each side to the eight-square (octagonal) lines.

Final trim—The eight-square yard was then rounded throughout its length except in way of the stops which remained four-square.

After the yard was trimmed to final shape iron hoops were driven on in way of the scarph. One wide hoop was placed at the slings and had an eye to receive the chain sling. One hoop was placed about nine inches inside of each nib of the scarph. The remainder were spaced about 30 inches apart.

Note: When the yard was slung from the masthead the scarph was always placed in the up-and-down position.

To Make a Topsail Yard from a Single Tree

The procedure for making a topsail yard is similar to that for producing a lower yard, differing only in minor details.

This yard, while considerably smaller than its companion lower yard, was a spar of considerable size and strength.

A topsail could be set, without reefing, in a wind category of "fresh breezes," which amounted to a wind pressure on the sails of almost one and one-half pounds per square foot.[39] In a large vessel this could amount to about 4,000 pounds acting against the sail and, ultimately, against the yard.

Following are the steps to produce a topsail yard.

First trim—Make a tree four-square and out of winding. Strike the middle line, quarters and yardarms.

Second trim—Mark the proportional diameters on the quarters and yardarms. Lay off stops from four feet to eight feet inboard of the yardarms. (This distance is based on contemporary masting tables and varies with the size of the vessel.) Trim to these lines. All sections through the stick are now rectangular except the slings. See Fig. 8.3.

Third trim—Cant the stick 90 degrees and repeat the second trim. All sections are now four-square.

Fourth trim—Line to eight-square by measuring 5⁄24 of each quarter from each side of the middle line and connect these points. This applies to all surfaces. Trim to octagonal section except the stops which remain four-square.

Final trim—Convert eight-square to round, except at the slings which retain the eight-square for one-eighth the length of the yard on each side. Retain four-square at the stops. Twelve inches in from the end of the yard cut a vertical sheave hole for the topsail reef tackle. Round off the end of the yard.

These details are illustrated in Fig. 8.3.

To Make a Topsail Yard of Two Trees and Two Fishes

When a large (over 12 inches in diameter) yard was required and a suitable tree was not available contemporary sources[39] record two methods of lengthening the yard. One method was to cut a long scarph, about ten feet, in the end of the yard and add the required length with another tree scarphed to suit. The joint was then reinforced with iron hoops.

The second method was to unite two trees with a scarph at the slings. The length of the scarph was two quarters of the length of the yard on each side of the slings. This joint would be reinforced by two fishes, one on each side of the

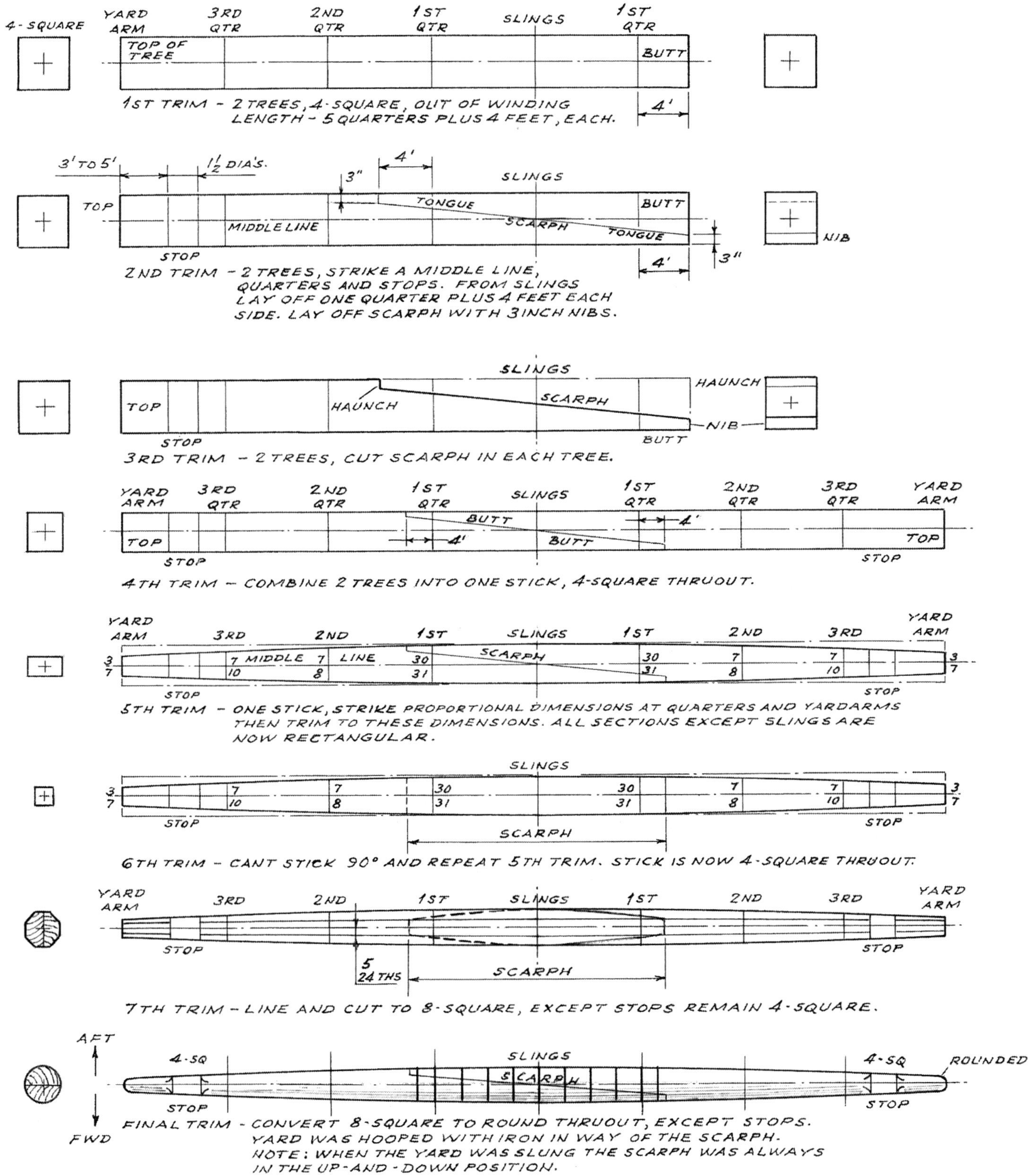

Figure 8.2. A lower yard made from two trees, scarphed.

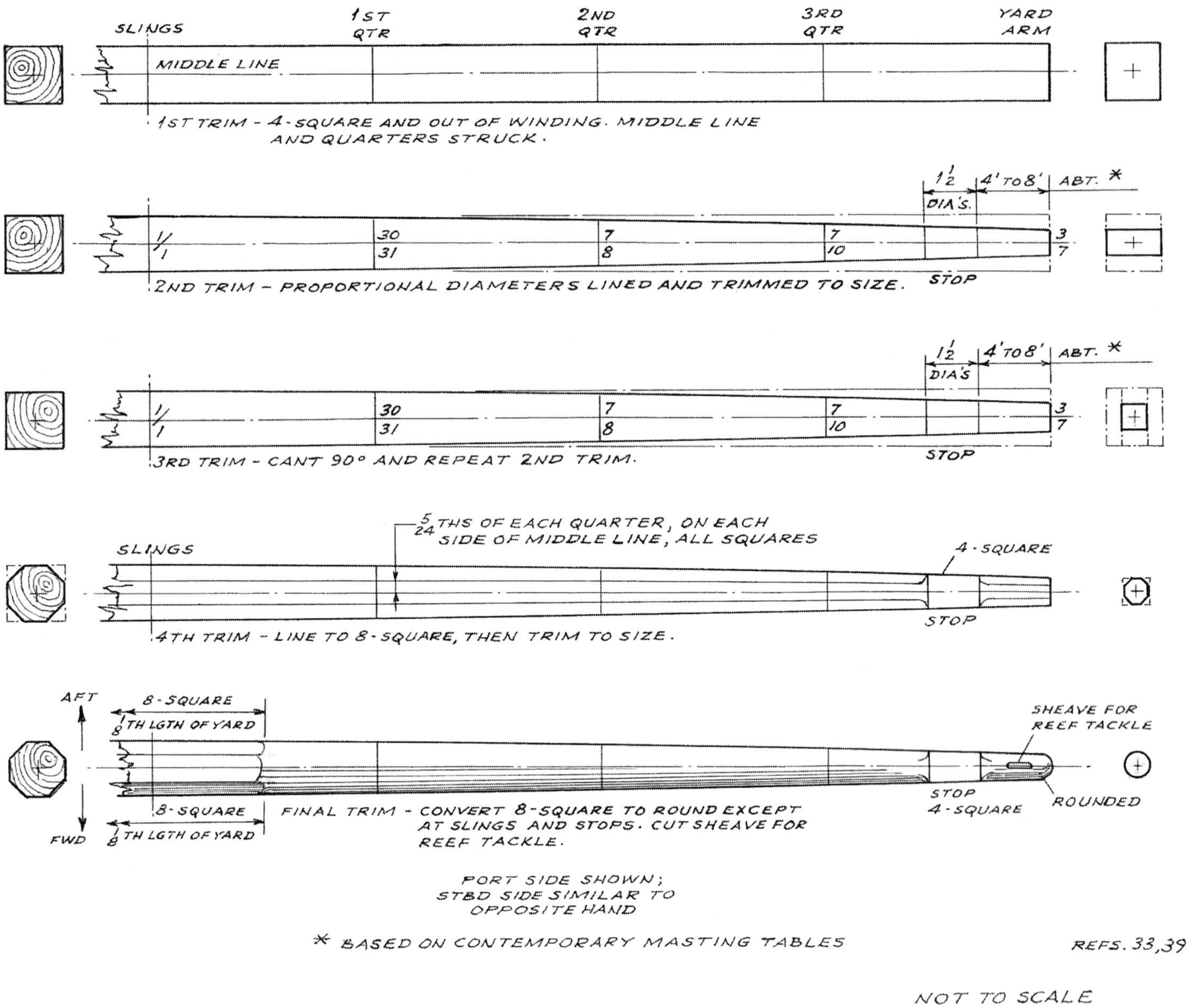

Figure 8.3. A topsail yard made from a single tree.

yard. The fishes were placed so that they covered the exposed line of the scarph. See detail in Fig. 8.4. This assembly, in way of the scarph, was hooped with iron for its entire length. The hoops were about 30 inches apart.

The weakness of the first method was that the scarph was in way of the most highly stressed portion of the yard. The second method was much stronger due to the location of the scarph and addition of the fishes.[39] This second method is illustrated in Figs. 8.4 and 8.5 which shows finished views and significant cross sections of this yard.

The construction sequence was as follows:

First trim—Two trees are made four-square and out of winding. Each tree is, in length, six quarters of the finished length of the yard. Line the slings, quarters and yardarms across the surface of the trees. Strike a middle line along the length of the surface.

Second trim—Two trees. Lay off the proportional dimensions at each quarter and yardarm. Connect these points. Lay off the locations of the stops. Lay off the scarph from 2nd quarter to 2nd quarter. Each scarph has a one and one-half inch tongue. This is an arbitrary figure.

Third trim—Two trees. Cut the scarph in each tree.

Fourth trim—Join the scarphs. Combine the two trees into one stick.

Fifth trim—With the scarph lying horizontally line the fishes on each side. Trim the trees to these new lines.

Sixth trim—Trim the fishes to fay with the trees. Attach the fishes and trim them to four-square. Install the ekeing in way of the fishes.

Seventh trim—Line all sides with proportional dimensions at each quarter. Trim four-square to these dimensions.

Eighth trim—Line all sides to eight-square by measuring 5/24 of the proportional diameter from the middle line on the slings, each quarter and the yardarms, except the stops remain four-square. Trim the yard to octagonal form on these lines, except the stops.

Final trim—Convert the eight-square to round throughout, except the stops. Cut a hole in each yardarm for the reef tackle sheave. Round the ends of the yardarm.

After the yard was trimmed to final shape iron hoops were driven on in way of the scarph and fishes. One hoop, with an eyebolt for the topsail tye, was placed at the slings and one hoop was placed about nine inches within the ends of the fishes. Other hoops were spaced about 30 inches apart along the fishes.

This yard, when slung on board, was hung with the fishes in a vertical position.

Fittings for the Lower Yards

After the trees had been converted from their original condition to the final form of the yard they required certain hardware before they could be slung on board ship.

Along the top of the yard a jackstay was fitted, port and starboard. This was iron, about ¾ inch in diameter, and was rove through iron eyebolts which had been driven into the yard at intervals of about 30 inches. The jackstay reached from inside the stops to the sling band on each side of the slings.

Hard against the outer end of the stops another iron band was fitted. This was made with three eyes—one at the top, one at the bottom and one on the aft side of the yard. They served the lifts, the reef tackles and the brace pendants, respectively. The diameter of the yardarm was sometimes let-down to provide a shoulder for this band to seat against.

On the crojack the middle eye was fitted on the forward side of the yard.

This band is shown in Fig. 8.6.

On the aft side of the stops, port and step, a cheek block was installed, through which the topsail sheets led.

At the yardarm the ironwork to accommodate the studdingsails was fitted. It consisted of a strap with long arms which passed around the end of the yardarm, two hoops to fix the strap in position, a crank and a ring. The ring had a lignumvitae roller inserted in its lower edge to ease the task of running out the studdingsail boom. The crank and ring were canted forward at an angle above horizontal. This is illustrated in Fig. 8.6.

At a distance of 3/16 × the length of the yard, in from the yardarm, another iron assembly called the quarter-iron was installed. This consisted of a ring pinned in position on the yard, and a second ring, supported by a stem, that occupied the same relative position as the ring on the crank. This ring supported the inboard end of the studdingsail boom, and was hinged about 30 degrees from the bottom and pinned at the top at a corresponding angle in the opposite direction. This allowed the studdingsail boom to be lifted at the inner end. The location of the hinge also prevented the boom from rolling out of the ring in a seaway. This is shown in Fig. 8.6.

This arrangement did not apply to the crojack, which was capped with a simple iron ferrule.

Masting tables for the packets and freighters of this period contain little, if any, reference to studdingsails while it is a safe assumption that all clippers, due to their mission of sailing swiftly, were outfitted with these sails.

In addition to ironwork that was installed exclusively on the lower yard there was also the truss, or cranse iron, which assembly provided the physical connection between the yard and the mast.

The first function of the truss was to support the yard at the proper height on the mast. Once this was achieved it was probable that the yard might never be removed from this location.

The second function was to hold the yard far enough away from the mast that, when braced hard up, the yard did not touch the shrouds and was clear of the stay.

Its construction allowed the yard to swivel horizontally when sailing conditions required that it be braced to catch a prevailing wind.

This truss was a formidable array of ironwork and it is reported that, in the clipper ship *Phantom*, the lower yards were held seven feet from the mast.

Basically, it consisted of the following parts: a hoop around the mast, a yoke, a universal joint, and two hoops (sometimes only one) around the yard. These parts were arranged in several fashions, three of which are illustrated in Fig. 8.7.

Fittings for Topsail Yards

After the topsail yard, whether it be a single tree or a made-yard, was tooled to final configuration, it required certain fittings to be installed prior to placement on board ship.

The first of these fittings was a sling band which was installed as a hoop made of iron and of a width equal to one diameter of the yard on which it was to be installed. It was placed at the mid-point of the yard which was called the slings. On the top of this sling band was a large eyebolt to which the topsail tye would be attached.

Next was the parral, the largest of the fittings. The parral was to the topsail yard what the truss was to the lower yard. However, where the truss maintained the yard in a permanent location, the parral was required to maintain the topsail yard in various locations on the mast.

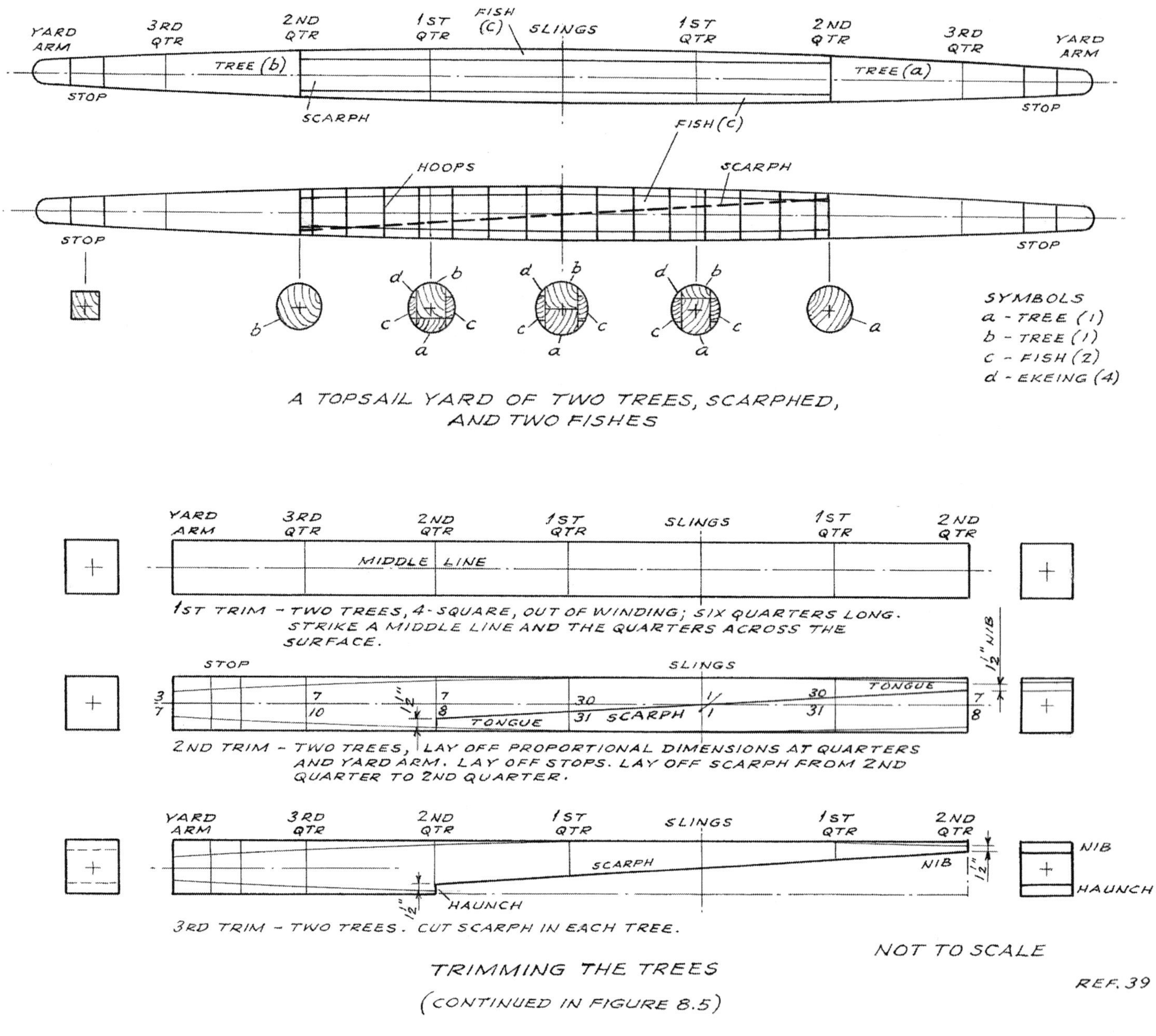

Figure 8.4. A topsail yard made of two trees, scarphed, and two fishes.

The typical parral consisted of a saddle that was secured to the middle of the yard. This saddle had a half-round cutout at one edge to which a half-round hinged iron strap with a toggle was attached. The resulting opening was of a diameter that, when lined with leather and allowing ¾ inch clearance all around, would allow the yard to move easily along the mast. The parral was secured to the yard with iron hoops and is illustrated in Fig. 8.6.

Also required, and installed at the outer end of each stop, was an iron band with three iron eyes attached. The upper eye was for the topping lift, the after eye was for the yard brace, and the lower eye was for the reef tackle.

On the mizzen yards the eye for the braces was located on the forward side of the band. This band is shown in Fig. 8.6.

Jackstays, iron, about ¾ inch in diameter, were secured along the top of the yard by means of eyebolts placed from the stop to the slings. These are shown in Fig. 8.6.

Studdingsail fittings consisting of a yardarm iron and a quarter-iron are shown in Fig. 8.6. The yardarm iron was made up of straps, a crank and a ring. The quarter-iron consisted of a complete ring, a stem, and a hinged ring.

Finally, a cheek block was secured to the aft side of the four-square stop for the topgallant sheet, and is shown in Fig. 8.6.

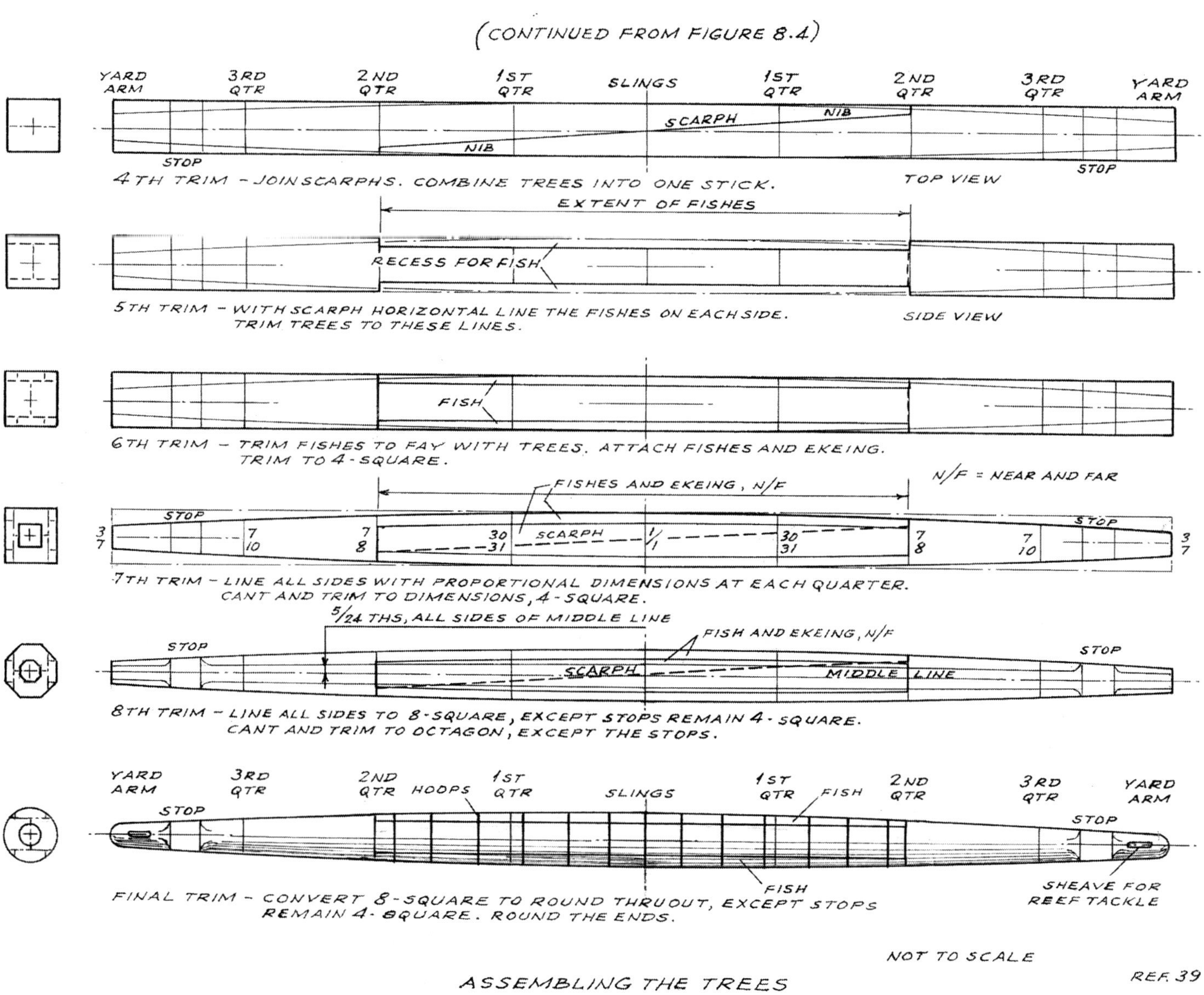

Figure 8.5. A topsail yard made of two trees, scarphed, and two fishes.

To Make the Upper Yards

It would be redundant to make drawings of the upper yards inasmuch as they offer nothing new or unique in construction or details. With this in mind the general details that apply and are contained in the various references are recorded herein.

Like topsail yards, lower yards and masts, these upper yards had their beginnings as four-square sticks out of winding. From that point they went through the same evolutions as their larger counterparts.

All were sized in proportion to the yards below. In some cases the lower yards were used as the reference base and in other cases the topsail yards were the basis. One source,[15] using the lower yards, quotes the lengths of the upper yards as follows: topgallant yard, 6/10 of the lower yard in length; royal yard, 5/10 of the lower yard; skysail yard, 3/10 of the lower yard.

Table 4.1 lists the topgallant yard as 3/5 the length of the topsail yard; royal yard, 1/2 the topsail yard; skysail yard (taken from actual vessels), about 4/10 the topsail yard; moonsail yard (taken from actual vessels), 3/10 and 4/10 the topsail yard.

In actual practice, based on the Ship Masting Tables listed in Chapter 4, the lengths of the yards varied widely but could be said to follow the values listed below. Compared to the length of lower yards the proportions were, for topgallant yards between 52 percent and 60 percent; for royal yards between 40 percent and 47 percent; and skysail yards between 25 percent and 35 percent.

Compared with topsail yards, topgallant yards were 66 percent to 78 percent as long; royal yards were 48 percent to

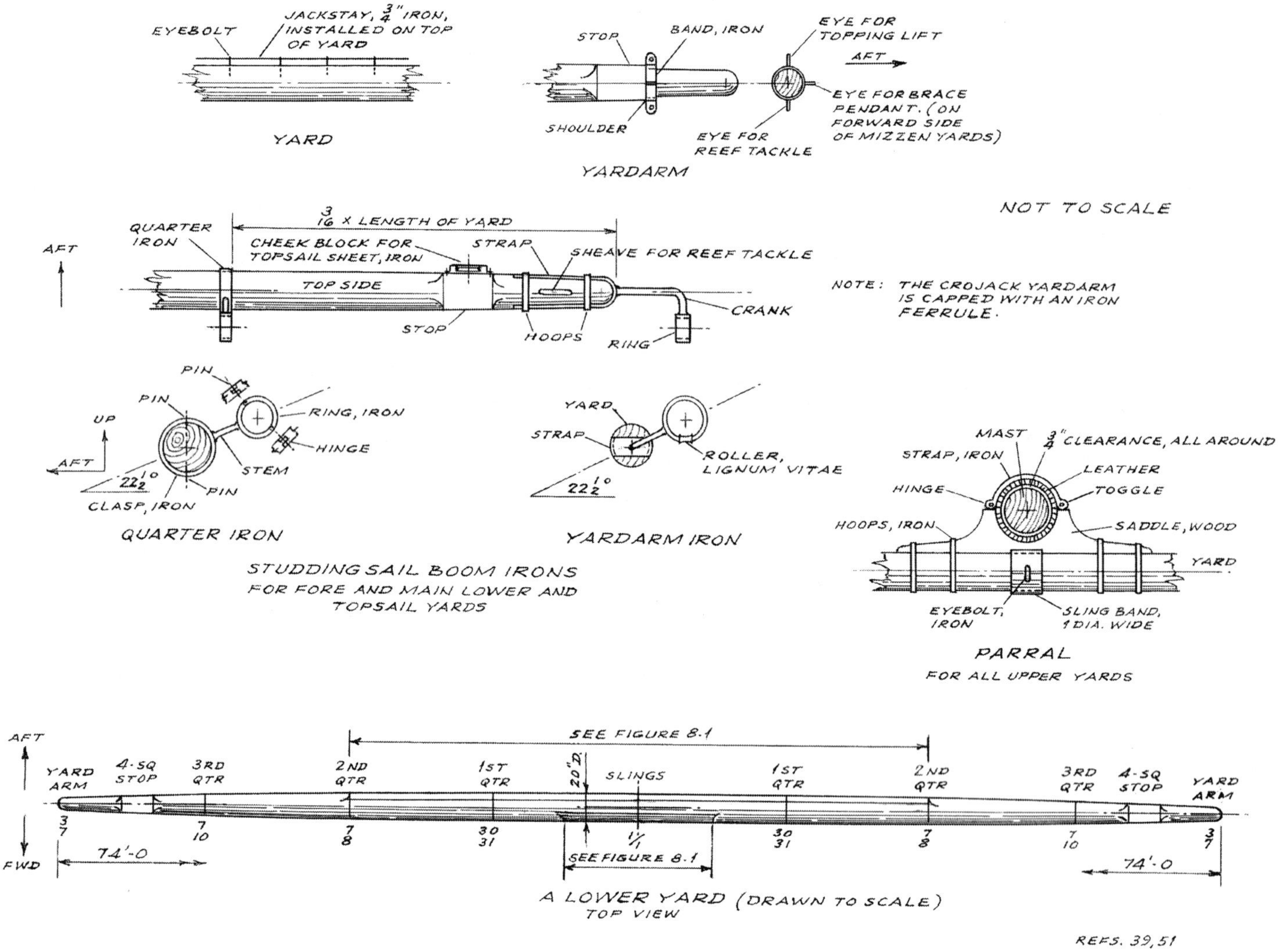

Figure 8.6. Fittings for fore and main lower and topsail yards.

56 percent; and skysail yards were between 33 percent and 43 percent.

Proportions of the diameter of spars to their length are contained in Table 4.2.[33]

The profiles of all these upper yards are the same as for the lower and topsail yards. The fractional proportions for each quarter and yardarm are given in Table 4.4. Actual measured diameters, at the quarters, for yards 22 inches in the slings to 3½ inches in the slings are given in Table 4.8.[33]

The smallest skysail yards found in this book are in the clipper ships *Flying Dutchman*, *Herald of the Morning*, *Wild Pigeon*, and *Witch of the Wave*, 1851. All are 3½ inches in diameter.

The Topgallant Yard

Topgallant yards were round throughout their length except the middle in way of the slings was left eight-square for ¼ the length of the yard, and the stops at the yardarm were left four-square. A sheave hole for the royal sheets was cut vertically through each stop. Immediately outboard of each stop an iron spider band with eyes for brace pendant and topping lift was installed. A jackstay was fitted along the top of the yard from the slings to the stop. See details in Fig. 8.6. Also required was a parral to confine the yard to the mast, and a sling band with an eye for attaching the topgallant tye. Both are shown in Fig. 8.6.

In ships that would carry royal studdingsails boom irons similar to those shown in Fig. 8.6 were fitted. In ships without such studdingsails the ends of the yard were capped with a ferrule and a jewel block for the topgallant studdingsail yard.[15]

These yards were trimmed to the fractional proportions tabulated in Tables 4.4 and 4.8.

The Royal Yard

Royal yards were round throughout their length except the middle in way of the slings was left 16- square for ¼ the length of the yard, and the stops at the yardarms were left four-square.

If the ship was fitted with skysails a vertical sheave hole for the skysail sheets was cut through each stop. A parral, a sling band, jackstay and brace bands at the yardarms were required similar to the topgallant yard and are shown in Fig. 8.6.

Royal yards were trimmed to the fractional proportions tabulated in Tables 4.4 and 4.8.

The Skysail Yard

In the absence of specific descriptions it appears that the skysail yards were made and outfitted in the same manner as the royal yards except that no sheave hole was required through the stops unless the vessel carried moonsails.

The Moonsail Yard

The moonsail yard was similar to the skysail yard except that its proportions were smaller. No sheave holes were bored through the stops. They were outfitted in the same way as the royal yards.

Moonsails were rare. Only three ships listed herein were fitted with them, they being *Young America*, *Hurricane* and *Seaman's Bride*.

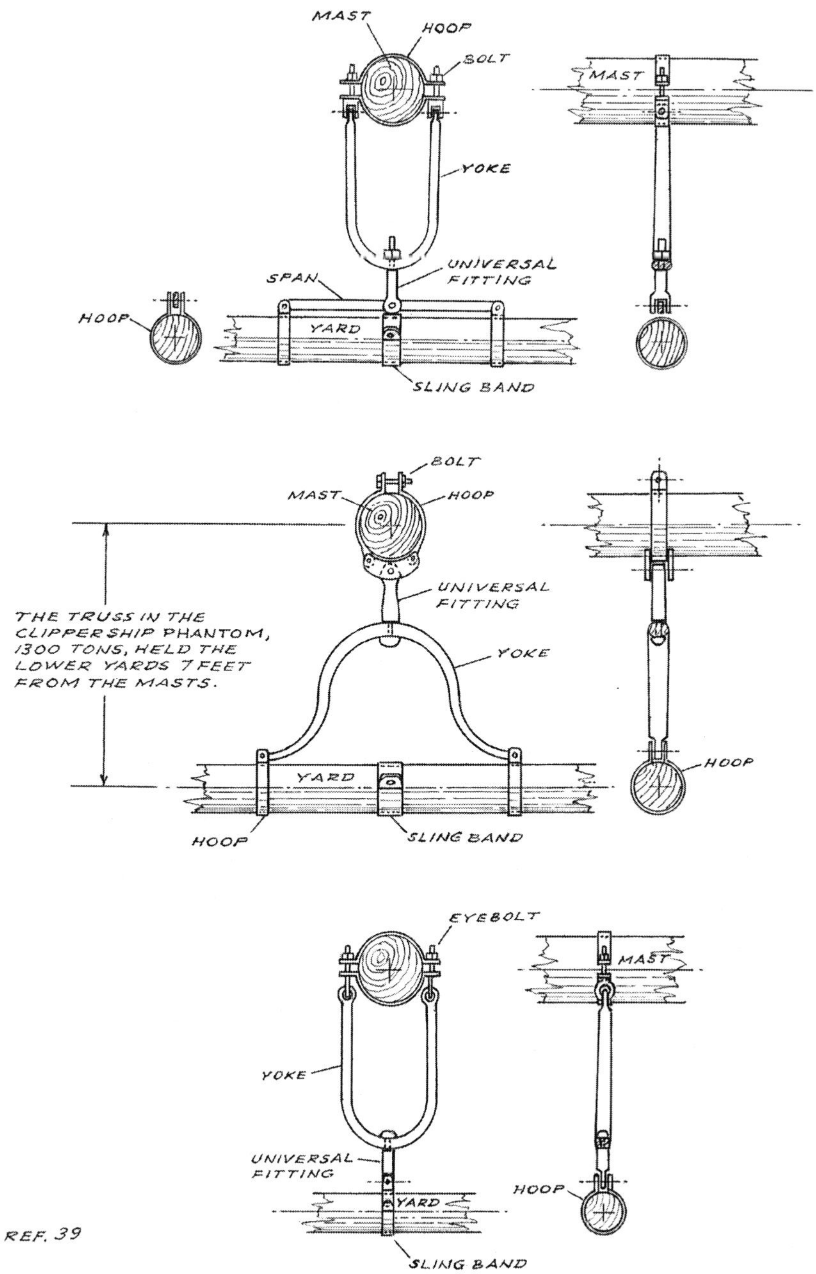

Figure 8.7. Three styles of lower yard truss.

9

Booms and Gaffs

The jibboom and flying jibboom built as a single stick was the most popular arrangement in the 1850s and later, but was predated by the use of a jibboom and flying jibboom as separate entities. The latter combination, however, was still very much in evidence during this time. Details of both arrangements are shown in Fig. 9.1.

To Make a Jibboom

Like all other masts and spars to be found on board ship the manufacture of the jibboom began with a stick four-square and out of winding.

The first operation was to strike a middle line along a surface of the stick then augment this by adding the controlling sectional lines across this surface. In the case of the jibboom the given or greatest diameter of the spar began at the heel and remained uniform until it reached the bowsprit cap. Between this point (the cap) and the outer end of the boom the length was divided into quarters. Each quarter was assigned a proportional value of the given diameter. These values are given in Tables 4.4 and 4.9

After these fractions were connected by a fair line the stick was sawed or hewed to this configuration. Then, as with all other spars, the jibboom was alternately canted, lined, hewed and converted to eight-square, finally to be rounded to the desired diameters, except at each end.

Both the heel and the outer end remained eight-square for three diameters length. At the heel a horizontal hole was bored through the middle eight-square for a sheave for the heelrope that hauled the jibboom out. Near the bitter end was a smaller round hole drilled through horizontally to accommodate a lashing to tie down the boom.

At the outer end, starting at the end and working aft within the length of the eight-square, were a half-sheave cut vertically; the flying jibboom iron, which was seated solidly on an eight-square necking; then a vertical sheave. Additional sheaves might be required depending upon the specific rig of the ship which carried this jibboom.

In addition, at some point about midway between the heel of the jibboom and the bowsprit cap, a hinged bale was secured around the bowsprit. The hinged upper portion restrained the jibboom in the desired position. Details of the above appear in Fig. 9.1.

A wooden saddle was fayed upon the bowsprit to insure parallel alignment of the jibboom with the bowsprit. This was covered in Chapter 5.

To Make a Flying Jibboom

The flying jibboom was processed in the same manner as the jibboom, however the final shape and proportions were decidedly different. The flying jibboom was about $\frac{2}{3}$ the diameter of the jibboom.

The heel of the flying jibboom was stepped in the bowsprit cap and was $\frac{3}{4}$ of the given or maximum diameter which was at the flying jibboom iron that was affixed to the forward end of the jibboom. The forward end of the flying jibboom was $\frac{2}{3}$ of the given diameter. Its length outside of the iron at the end of the jibboom varied widely between $\frac{2}{3}$ and $\frac{7}{8}$ of the inner length. This variation is based on examination of actual ship masting tables.

The fractional diameters at the quarters were as follows. The given diameter was at the end of the jibboom (the flying jibboom iron). This diameter was maintained aft for $\frac{1}{3}$ the distance to the bowsprit cap. Quarters were marked off between this point and the heel, and diameters established as in Table 4.4. The same procedure was applied between the flying jibboom iron and the forward end of the boom.

The final shape of the flying jibboom was round for its entire length except for three diameters at the heel which remained eight-square. A horizontal hole was bored through the middle eight-squares to accept a lashing to restrain the boom. The forward end had a necking about one diameter aft, and then a vertical sheave.

If necessary the bowsprit cap was made full on the starboard side to provide a step for the flying jibboom. These details are shown in Fig. 9.1.

To Make a Jibboom and Flying Jibboom in One

An inspection of the details applying to a jibboom and flying jibboom being shipped separately leads to the inevitable conclusion that the two could be economically combined into a single stick. In addition, the result had the advantage of simplicity. These characteristics are shown in Fig. 9.1.

The development of the original four-square stick was the same as has been described for other spars with very minor changes.

As with a separate jibboom the heel and the length from there to the bowsprit cap was of uniform maximum diameter and detailed in similar fashion. The distance from the bowsprit cap to the stop at the forward end of the flying jibboom was divided into quarters having the usual proportional values. At the second quarter a stop at the forward end of the jibboom separated it from the flying jibboom. The fourth quarter marked the extreme end of the flying jibboom with a stop, beyond which a pole extended.

At each stop a spider band with appropriate eyebolts for securing various ropes was fixed. In addition, holes for sheaves that positioned the headstays were bored as required for individual vessels.

Details of this jibboom are shown in Fig. 9.1.

Miscellaneous Booms

As did all previous spars in this book the assorted array of booms and gaffs that follows had their beginnings in the same fashion. Each one began as a stick, four-square and out

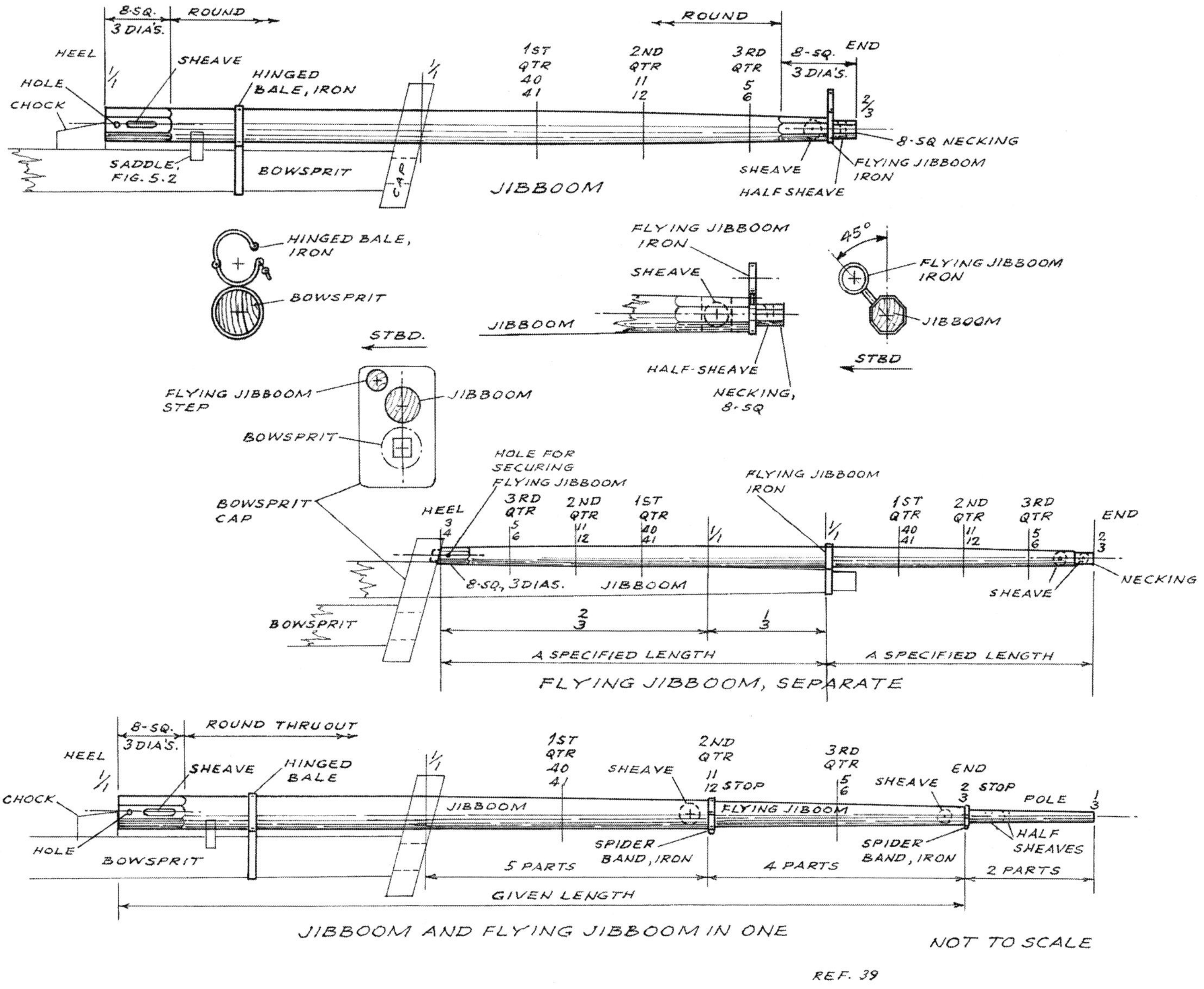

Figure 9.1. The jibboom and flying jibboom.

of winding. From that state the various stages of trim were developed—lining, hewing, canting, etc., until a round spar resulted. From that stage the final details for individual spars were applied to each stick as related to the specific function of the boom or gaff.

The Spanker Boom

Sources differ in assigning the location of the maximum, or given, diameter of the spanker boom. It is variously listed as being at mid-length of the boom or at the taffrail (about one-third of the total length from the aft end of the boom). In either case the quarters were marked off in both directions from the maximum diameter. The fractional values of the quarters are given in Tables 4.4 and 4.9. See also Fig. 4.1.

The butt end of the tree was located at the after end of the boom.

The tree was converted to round for its entire length except at the inner end which was left square for four or five feet to allow for the jaws to be worked into the stick. The jaws, in thickness, were about one inch less than the diameter of the inner end.

The end of the boom was formed into a tongue that scarphed into the jaws, which were made of oak, and formed to a half circle one inch larger than the diameter of the mast.

This allowance was for leather and play. The jaws were then trimmed to an inward curve to fair with the size of the boom, leaving sufficient strength to the hollow of the jaws. The forward ends of the jaws were rounded.

The jaws were secured to the boom by three or four iron hoops driven into position.

The outer end of the boom was trimmed to a short, square necking. A vertical hole for a sheave was cut through the necking.

Eyes for the inner and outer topping lift pendants were generally secured with iron hoops.[33]

See Figs 4.1 and 9.2.

The Mainsail Boom (Schooners)

This boom was lined and converted to round in the same manner as the spanker boom, and it too maintained a four-square for about four feet at the inner end to accommodate the jaws.

The general configuration of the two booms was similar. However, the fractional values of the intermediate quarters differed. They are listed in Tables 4.4 and 4.10 for the mainsail boom. See also Figs. 4.1 and 9.2.

The Lower (Swinging) Studdingsail Boom

The given diameter of this boom was at the heel and was maintained outward for one-third of the boom length. From this point it decreased uniformly in diameter until it was ⅔ of the given diameter at the end. (Existing tables listing the fractional diameters at the various quarters of ships' spars do not include values for studdingsail booms or yards.) See Fig. 4.1.

The booms were rounded throughout their entire length and had a square necking at the outer end, approximately one diameter in length. A vertical hole was drilled through the whole diameter immediately within.

The inner end was fitted with an iron ferrule and gooseneck. The proportions of this boom were one inch diameter for every five feet of length. See Fig. 4.1 and Table 4.2.

Topmast and Topgallant Studdingsail Booms

These booms were peculiar unto themselves, in that the given diameter extended along the entire middle third of the boom, from which point it tapered to the ends of the boom. They are rounded the entire length and have a hole in the inner end for a lashing.

The proportions of the booms were one inch of diameter for every five feet of length. See Fig. 4.1 and Table 4.2.

The Martingale Boom (Dolphin Striker)

Martingale booms were small spars which hung from the underside of the bowsprit cap. Their purpose was to keep the martingales and the martingale backropes in position to counteract the upward forces acting on the jibboom by the headstays and the headsails.

The booms were not governed by formula and their sizes varied greatly, depending largely on the size of the ship in which they were fitted. In general they reached down from the underside of the bowsprit cap for a distance of slightly more than halfway to the waterline.

In actual size they ranged from 10 feet to 17 feet long and seven inches to 10 inches in diameter at their mid-length. They tapered from the middle toward the ends in an ovate taper. At the upper end a hook was inserted and at the lower end iron bands and eyes were fitted to receive the martingales and the martingale backropes. The various headstays passed under cleats which were installed in the lower half of the boom. See Fig. 4.1.

Topmast, Topgallant and Royal Studdingsail Yards

These yards were, in length, about one-half the length of their respective booms. Their diameters were one inch for every five feet of length.

The given (maximum) diameter was at the mid-point of the yard. From this point they tapered toward each end where they were ¾ of the given diameter. See Table 4.2. The yards were round throughout their entire length.[15]

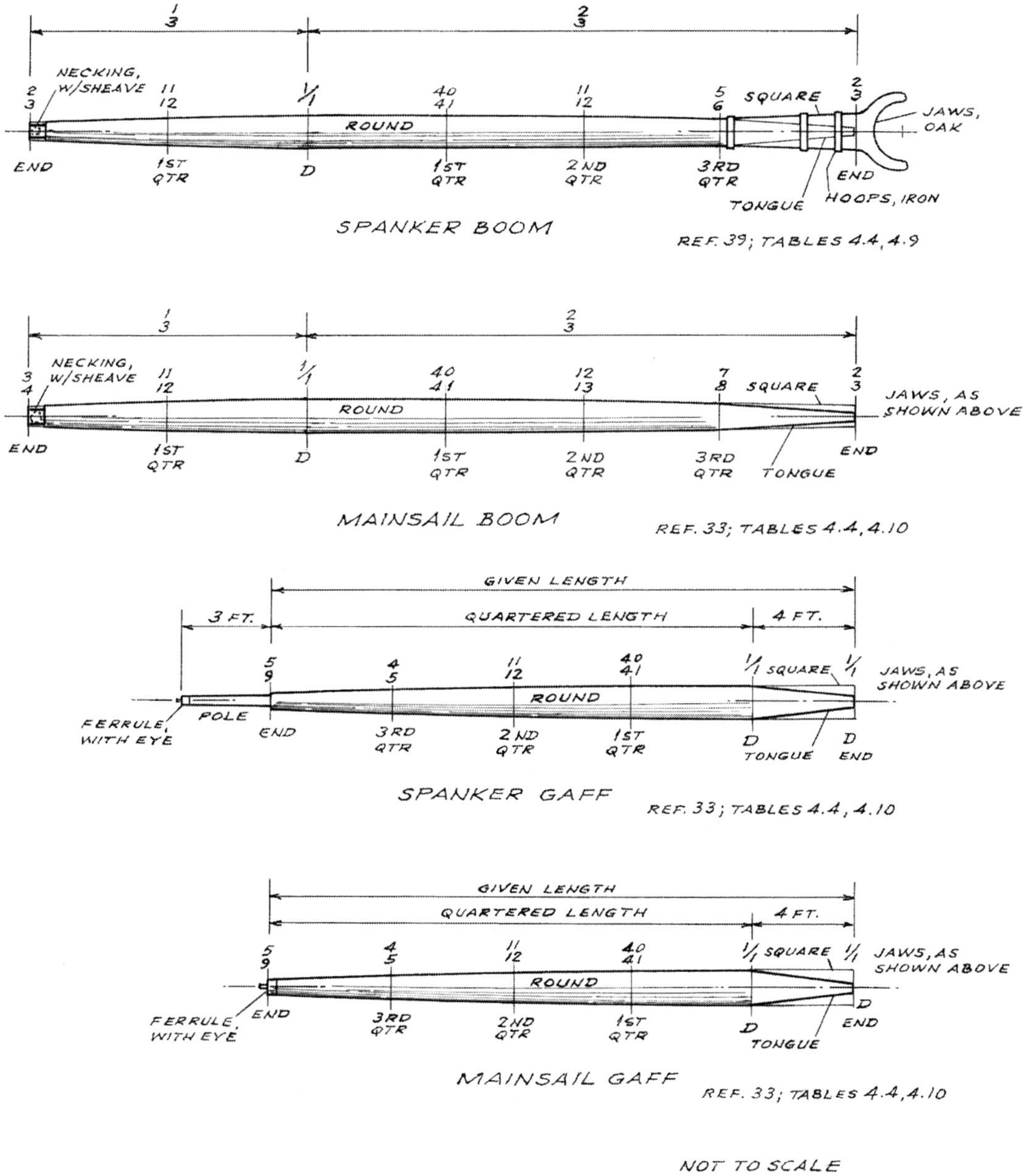

Figure 9.2. Proportions of spanker and mainsail booms and gaffs.

Miscellaneous Gaffs

The Spanker Gaff

The spanker gaff was converted from four-square to round except at the inner end where it remained square for about four feet to receive the jaws.

The length of the gaff was about 7⁄10 of the length of the spanker boom and was measured from the jaws to a stop at the outer end or peak. A round signal pole, about three feet long, extended beyond the stop. Its purpose was to prohibit any signal flags from being hidden by the sail.

The given length of the gaff was taken from the jaws to the stop. The overall length included the extension for the signal flags beyond the stop. The given diameter of the gaff was ⅞ inch for every three feet of length.

From the jaws to the stop it was divided as follows: the inner end was square for four feet; from this point out to the stop it was divided into quarters which were given proportional values as shown in Tables 4.4 and 4.9. The quarters were lined and made round. Outside of the stop the pole was made round. See Fig. 9.2.

Within the four-square end the gaff was shaped to a tongue which fayed with the scarph of the jaws which were shaped to suit the mast, with about one inch clearance. The hollow of the jaws was steeply beveled to allow the gaff to be properly peaked. The jaw assembly was united with the gaff tongue by three or four iron hoops driven on.

The extreme end of the pole was fitted with an iron ferrule and an eye, up and down, for a dasher block.

The Mainsail Gaff (Schooners)

The mainsail gaff was lined and sized in the usual fashion. Its length was about ⅗ of the length of the mainsail boom. Its diameter was ⅞ of an inch for every three feet of length.

The stick was converted to round for its entire length except for four to five feet at the inner end which portion remained four-square for fitting the jaws, which were fashioned in the same manner as those of the spanker gaff and hooped in the same way.

Hardware was installed to suit the rigging. The proportional diameters are taken from Tables 4.4 and 4.10. See Fig. 9.2.

The peak was fitted with an iron ferrule and an eye, up and down, for a dasher block.

The Trysail Gaff

The length of the trysail gaff was ⅝ that of the spanker boom of the ship in which it was fitted. The given diameter was at the inner end and was based on the length of the gaff. The value of the diameter was about ⅞ inch for every three feet of length. The diameter at the outer end was ⅗ to 5/9 of the given diameter. See Fig. 4.1 and Tables 4.3 and 4.4.

The gaff was round throughout its length except that about four feet at the inner end was left square for fitting the jaws which were steeply beveled in the hollow.

An iron ferrule and an eye, up and down, were fitted in the end for a block for displaying signals.

The Spencer Gaff

Spencer gaffs were finished round throughout their length, and they were generally made as standing gaffs fitted with an iron ferrule and a gooseneck at the inner end, and a short necking at the peak.

The entire length was divided into quarters and each quarter given a fractional value in accordance with Table 4.4.

Actual masting tables indicate wide variance in their size, ranging between 15 and 25 feet in length, with short ends, and from four to seven inches in diameter. See Fig. 4.1.

Spencer gaffs were fitted on fore and main masts only in square rigged vessels.

The Monkey Gaff

The monkey gaff was a small spar which was rigged as a standing gaff from the mizzen topmast crosstrees.

It was round throughout its length, which varied between 15 feet and 22 feet, according to existing masting tables, and was about five inches in diameter at the butt, tapering to about 2½ inches at the peak.

The butt end was fitted with an iron ferrule and a gooseneck while the peak had an iron ferrule and an eye, up and down, to receive a dasher block. The halyard for the national ensign was rove through this block. See Fig. 4.1.

The foregoing pages represent all the principal spars that were to be found in the American merchant fleet of the 1850s.

Part II

Masting Arrangements

10

Location and Rake of Masts

No phase of wooden sailing ship construction was more subject to variables than the placement of masts in the hull. The height of masts, the spread of canvas, the configuration of her lines under water, were all of significant importance in positioning the masts so that the vessel could achieve her maximum sailing capabilities.

Naval vessels, being designed and built to fit into a definite rate or class, benefited from the findings of previous vessels of a given rate or class. Because of their similarities it was possible to compile tables that allowed one vessel to be copied from another. A benefit of this was that the installation of masts in their optimum locations was pre-determined.

In merchant fleets there was no such establishment to dictate the terms under which a ship could be built. There were certain restrictions, such as those ordained by Lloyd's and, at a later date, the American Bureau of Shipping, but in general a ship was built in accordance with the owner's desires and the designer's knowledge.

The spacing of the masts varied with the form of the ship and it was evident that a full-bowed vessel allowed her foremast to be stepped further forward than a sharp one. Though a general rule might be espoused, it was essential that every designer or builder should consider the nature of the form of his particular vessel and vary the disposition of her masts accordingly.[33]

The length on the waterline was the basis for determining the location of the masts. The mid-point of this length was the point from which the exact location of each mast was measured.

Fincham[39] states, "Calculations made by constructors who, in general, rest satisfied with making a comparison with other ships, and placing the point of sail according to their judgement of the form of the body."

In addition he observes that "in some American ships the foremast has been placed .34 before the middle of the waterline, and the mainmast .10 the length of the waterline aft of the middle. Mainmasts have been placed from .05 to .079 abaft the middle of the waterline length."

While no tables were prepared for use in masting American vessels, the descriptions of many such ships contain the actual dimensions of mast spacing.

Table 10.1 records the spacing of masts, where known, for vessels covered in this book. The spacing of masts in a ship refers to the distance that one mast is placed in reference to the next nearest, as the foremast from the mainmast, and the mainmast from the mizzenmast. These distances are variable depending upon the size of the vessel and the philosophy of the designer and are, in general, governed by the length of the yards that will be slung from each mast. For best service sails were made to sizes that would not conflict with each other when close hauled. One sail should not prevent a wind from acting on another sail.

The location of each mast refers to the actual spot on the deck where the mast was to be stepped. This information is covered in Table 10.2. However, due to the lack of information as to the waterline lengths of the ships involved the proportions are based on the length of the deck in these ships. All entries in Table 10.2 are taken from the mid-point of the ships' length.

In all cases the mainmast is stepped close aft of the mid-point of the ship.

The rake of the masts, more than any other feature of a ship's design, evokes the conception of speed. The reality is that, when the masts are raked, the wind has less power on the sails.

The wind, blowing in a horizontal direction, strikes a sail set on a mast with no rake, at a right angle, or square against the wind. This single horizontal force presses against the surface of the sail and expends itself as the ship moves forward.

A wind blowing in a horizontal direction, striking a sail that is set on a mast with rake, presses the sloping surface of the sail and is divided into two components, namely, the original horizontal force and a resulting vertical force. The vertical component tends to push the sail up but, at the same time, diminishes the horizontal force thus reducing the forward motion of the ship.

The following is quoted from Kipping, "When a sharp-bowed vessel has her masts to rake, it frequently eases her in pitching, but never adds to her sailing, the wind having less power on her sails."[33]

Table 10.3 lists the ships included in this book that had

Table 10.1.
Spacing of Masts

Name of Vessel	*Foremast Aft of Stem*	*Foremast to Mainmast*	*Mainmast to Mizzenmast*	*Mizzenmast to Taffrail*
Alice Tainter, bark 140 ft. o.d.	28½ ft. aft of stem	53 ft.	33½ ft.	27 ft.
Challenge 252 ft. o.a.	54½ ft. aft of knightheads	75½ ft.	56½ ft.	43'-3" to rudder stock
Constantine 205 ft. o.a.	39 ft.	66 ft.	49 ft.	34 ft. to sternpost
Flying Cloud 235 ft. o.a.	58 ft. aft of knightheads	69 ft.	54 ft.	44 ft.
Flying Fish 220 ft. o.a.	52 ft.	62 ft.	52 ft.	44 ft. to sternpost
Great Republic, 1853 335 ft. o.a.	76 ft.	103 ft.	85 ft.	mizzenmast to spanker 46 ft., spanker to taffrail 25 ft.
Great Republic, 1855 325 feet o.a., rebuilt	70 ft.	103 ft.	85 ft.	mizzenmast to spanker 46 ft., spanker to taffrail 21 ft.
Hurricane 230 ft. o.a.	52½ ft. aft of knightheads	72½ ft.	57½ ft.	45 ft.
Joseph Walker 180 ft. o.d.	36 ft.	64 ft.	42½ ft.	33 ft.
Josephine, brigantine 130 ft. o.d.	27 ft.	58½ ft.	45 ft. to rudder post	
Lightning 243 ft. o.a.	48 ft.	75 ft.	66½ ft.	42 ft. to rudder stock
Manhasset, schooner 101 ft. o.d.	26 ft.	35 ft.	40 ft. to taffrail	
Ocean Monarch 242 p.p.	54 ft. aft of knightheads	78 ft.	58 ft.	51 ft.
Plandome, schooner 101 ft. o.d.	26 ft.	35 ft.	40 ft. to taffrail	
Racehorse, bark 128 ft. o.a.	29½ ft. aft of knightheads	49 ft.	28½ ft.	21'-3"
Red Jacket 242 ft. o.a.	50 ft. aft of knightheads	78 ft.	64 ft.	49'6"
Sea Witch 192 ft. o.a.	39 ft. aft of knightheads	57 ft.	47 ft.	34'-6"
Stag Hound 226 ft. o.a.	50 ft.	67 ft.	56 ft.	42 ft. to sternpost
Surprise 190 ft. o.a.	36 ft.	59 ft.	53 ft.	36 ft. to sternpost
Trieste, bark 135 ft. o.d.	26 ft.	51 ft.	35'6"	24 ft.
Volante, brig 112 ft. o.d.	29'6"	41 ft.	42 ft. to taffrail	
Wild Pigeon 184 ft. o.a.	37'8" aft of inner stem	53'6"	44'6"	33'-6" to sternpost
Witch of the Wave, 1851 220 ft. o.a.	45 ft.	67 ft.	53 ft.	37 ft.

Table 10.2.
Location of Masts

Based on percentage of vessel's length on Waterline

Name of Vessel	*Length of Vessel*	*Foremast forward of midpoint*	*Mainmast aft of midpoint*	*Mizzenmast aft of midpoint*
Alice Tainter, bark	147 ft.	.303	.077	.310
Challenge	230	.261	.065	.304
Constantine	188	.287	.064	.319
Flying Cloud	225	.253	.053	.293
Flying Fish	210	.248	.052	.300
Great Republic, 1853	335	.278	.030	.284 Spanker aft of midpoint .421
Great Republic, 1855	325*	.278	.030	.284 Spanker aft of midpoint .421
Hurricane	223	.272	.053	.303
Joseph Walker	176	.295	.068	.318
Josephine, brigantine	130	.300	.138	—
Lightning	232	.293	.026	.302
Manhasset, schooner**	101	.248	.100	—
Ocean Monarch	241	.274	.050	.290
Plandome, schooner**	101	.248	.100	—
Racehorse, bark	124	.258	.145	.363
Red Jacket	242	.293	.029	.298
Sea Witch	178	.281	.039	.298
Stag Hound	216	.269	.042	.301
Surprise	184	.310	.011	.288
Trieste, bark	136	.309	.066	.331
Volante, brig	112	.241	.125	—
Wild Pigeon	170	.270	.053	.312
Witch of the Wave, 1851	202	.272	.059	.317

*Length on waterline did not change. **These schooners were built from the same plans.

Table 10.3.
Rake of Masts

In inches per foot

Name of Vessel	*Foremast*	*Mainmast*	*Mizzenmast*	*Spanker mast*
Abbott Lawrence	¾	⅞	1¼	
Alice Tainter, bark	¾	⅞	1¼	
Belle, brig, 1850	1	1	—	
Benin, bark	1	1¼	1½	
Black Hawk	¾	⅞	1⅛	
Challenge	1⅛	1¼	1½	
Champion of the Seas	½	⅝	1	
Constantine	¾	1 1/16	1½	
Dauntless	1	1	1¼	
Ellen Foster	1¼	1½	1¾	
Elvira	⅞	1	1¼	
Empress of the Sea	⅞	1	1⅛	
Flying Cloud	1¼	1¼	1¼	
Flying Fish	1¼	1¼	1¼	
Game Cock	1	1⅛	1¼	
Great Republic, 1853	½	½	½	1½
Great Republic, 1853 (rebuilt)	½	½	½	1½
Harry Bluff	½	⅝	1	

Name of Vessel	*Foremast*	*Mainmast*	*Mizzenmast*	*Spanker mast*
Henry Hill, bark	1⅛	1½	1⅞	Most extreme rake
Hesperus	⅞	1	1¼	
Hoogly	1	1¼	1½	
Invincible	1⅛	1¼	1½	
J. Montgomery	1	1⅛	1⅜	
John Bertram	1	1⅛	1¾	
Joseph Walker	¾	¾	1	
Josephine, brigantine	¾	1	—	
Lightning	¾	¾	1¼	
Manhasset, schooner	1¼	1¼	—	
Nightingale	1¼	1½	1¾	
Ocean Monarch	¾	1	1¼	
Plandome, schooner	1¼	1¼	—	
Polar Star	1	1¼	1½	
Queen of the Seas	1	1¼	1½	rake
Racehorse, bark	1½	1½	1½	
Racer	1	1¼	1½	
Red Jacket	1⅛	1¼	1½	
Sea Bird, bark	1¼	1½	1¾	
Sea Witch	1	1¼	1¼	
Shooting Star	1³⁄₁₆	1⅜	1¾	
Sovereign of the Seas	¾	⅞	1⅛	
Staffordshire	1¼	1¼	1¼	
Staghound	1¼	1¼	1¼	
Storm King	1	1¼	1½	
Sultana, bark	1	1¼	1½	
Surprise	⅝	¾	1	
Trieste, bark	1	1	1¼	
Volante, brig	¾	1	—	
Water Witch	1	1½	1¾	
Westward Ho	¾	⅞	1⅛	
Wild Pigeon	1¼	1⅜	1½	
Wildfire, bark	1¼	1½	1¾	
Winfield Scott	1	1¼	1½	
Witch of the Wave, 1851	1¼	1½	1¾	
Witch of the Wave, 1856	¾	⅞	1¼	
Witchcraft	1	1¼	1½	
Young Turk, bark	1	1¼	1½	

known rake to their masts. The least rake noted is ½ inch per foot, for foremasts, which applied to several vessels; the greatest rake noted is 1⅞ inches per foot, which applied to the mizzenmast of the bark *Henry Hill*. Whether or not all vessels had their masts to rake can only be conjecture.

However, it was axiomatic that all main and mizzenmasts should be raked. As explained by Kipping, "It is necessary that a ship's main and mizzenmast should rake more than the foremast, for, by separating them in this way, the wind acts with more power on all the sails, when close-hauled, which otherwise would not be effected, and be of little or no advantage to the ship."[33]

A final consideration in the stepping of masts in a ship involved the location of the mast in way of the partners. Steel[47] advises that care should be taken that a mast shall not touch the partners at any point since the mast might spring or be carried away.

Masts were made of soft wood, generally fir. If the mast should bear directly against the oak partners rather than against the wedges, the tangent point of the diameter of the mast would bear the entire load that was generated against the mast. This point of contact of the mast could easily crush or disfigure under such a load. When the mast was restrained by wedges the same load was distributed over a substantial area, thus reducing the unit pressure on the mast.

Wedges were carefully shaped and installed to fill the allowance of space between the masts and the partners.

11

Spar Plans of American Merchant Sail in the 1850s

A casual observer of the waterfront scene of any seaport in the 1850s would see, and be impressed by, the forest of masts and spars that was spread before him. Dozens of vessels berthed at the numerous piers presented a tangled mass of wood and rope. Nothing about the scene presented an orderly appearance.

A closer glimpse would reveal that this apparent entanglement was made up of many vessels of all rigs and sizes, each of which contributed to the unorganized scene but was, individually, a very precisely arranged entity.

A detailed inspection of any individual vessel would reveal a sleek hull, neatly aligned masts, generally two or three to each vessel, and a selection of yards aloft slung at various angles.

Most of the vessels that made up this panorama were large, three masted, square riggers. However, other rigs that were well known at the time might be found in the array.

There were six classes of vessels common in American seaports at this period, they being three mast ships, three mast barks, brigs, brigantines, two mast topsail schooners and two mast schooners. These formed the bulk of American merchant sail in the 1850s and all are represented among the vessels named in this book.

All in all, 75 specific spar plans have been discovered. In any given class of vessel the differences range between obvious and minute yet each is distinct in its own right.

These spar plans are illustrated by line drawings and are included in the following pages. Also included are brief descriptions of the details and names of the vessels which were fitted with the spar plan.

The vessels are listed in descending order starting with the largest and most complex arrangement of the square rig and concluding with the simplicity of a two mast schooner.

Seventy-five variations of masting and sparring are depicted in the illustrations. While this is a considerable total it cannot be assumed that it covers every conceivable spar plan. However, a close inspection of the sketches causes one to wonder what else could be added.

For all that has been written and researched about American wooden sailing vessels of the mid–19th century little time has been devoted to the detailed intricacies of mast and spar making.

Table 11.1 lists the names of all the ships included in this book and notes the number of the spar plan which is applicable to each vessel.

These imposing structures, some of which rose 200 feet above the water, were not subject to minute inspection, as was a ship's hull. In addition, their appearance, which was plain and simplistic, did little to excite interest. They were simply recognized as an integral element of the sailing ship.

Generally speaking, a curious person would be obliged to refer to original books published in the period of the vessels involved. As has been revealed in this book, the details are amazing.

Figure 11.1 illustrates a square rigged ship with single topsails. It is fitted with every basic spar that such a vessel could carry. All the spars shown can be found in the sparring sketches in some variation. However, no vessel included herein is fitted with all 44 spars shown.

One phase of sparring that has been handed down through time are sparring tables which contain the quantities and sizes of spars for given vessels. This chapter contains much of such information. All dimensions are taken from contemporary (1850s) sources.

Descriptions of Spar Plans

Illustrations of the spar plans follow these descriptions.

Note: All rigs, #1 through #72, are fitted with a spanker boom and spanker gaff.

Spar Plan #1—Four masts, bark; Forbes rig; three jibs in a single spar; skysails on fore, main and mizzen. Spanker topmast is a single pole.

Applies to *Great Republic* as built in 1853.

Spar Plan #2—Four masts, bark; Howes rig; three jibs in a single spar; skysails on fore, main and mizzen. Spanker topmast is a single pole.

Applies to *Great Republic* as rebuilt in 1855, after she was burned at her pier.

Table 11.1.
Masting Arrangements

Vessel	*Year*	*Spar Plan*	*Category*	*Rig*	*Remarks*
Abbott Lawrence	1855	53	packet	ship	
Alarm	1856	8	clipper	ship	Howes rig
Alice Tainter	1856	65	freighter	bark	
Alma	1855	51	freighter	ship	
Amos Lawrence	1855	53	packet	ship	
Amphitrite	1853	7	clipper	ship	Forbes rig
Andrew Jackson	1855	12	clipper	ship	Howes rig
Antarctic	1850	56	packet	ship	
Antelope	1855	69	freighter fitted with screw propeller	demi-barque	Barkentine classification did not appear until about 1870
Asa Eldridge	1856	8	clipper	ship	Howes rig
Aurelia	1855	59	freighter	bark	Howes rig
Aurora	1854	21	packet	ship	Forbes rig
Bald Eagle	1852	43	clipper	ship	
Barreda Brothers	1854	53	packet	ship	
Belle	1850	70	packet	brig	
Belle	1853	72	clipper	brig	Masts "Baltimore clipper raked."
Belle of the West	1853	6,7	clipper	ship	Upper masts fidded abaft lower masts. References to not agree. Forbes rig.
Benin	1854	65	freighter	bark	
Black Hawk	1856	46	clipper	ship	
Blue Jacket	1854	15	clipper	ship	Howes rig
Bounding Billow	1854	65	clipper	bark	
Brenda	1851	74	freighter	topsail schooner	
Cathedral	1855	53	packet	ship	
Challenge	1851	29	clipper	ship	
Challenger	1853	34	clipper	ship	
Champion of the Seas	1854	49	clipper	ship	
Charmer	1854	36	clipper	ship	
Chatsworth	1854	53	packet	ship	
Clara Wheeler	1850	49	packet	ship	
Climax	1853	13	clipper	ship	Howes rig (introduced)
Coeur de Lion	1854	33	clipper	ship	
Comet	1851	28	clipper	ship	
Commodore Perry	1854	20	packet	ship	Howes rig. Fitted with sliding gunter royals.
Constantine	1850	49	packet	ship	
Cornelius Grinnell	1850	16	packet	ship	Forbes rig (original). Upper masts fidded abaft lower masts.
Dauntless	1852	34	clipper	ship	Howes rig. Spar plan probable.
Dione	1857	24	freighter	ship	Lower topsail yards are stationary, slung with lifts.
Donald McKay	1855	11	clipper packet	ship	Howes rig
Eagle Wing	1853	43	clipper	ship	
Edisto	1851	67	packet	bark	
Edwin Forrest	1853	34	clipper	ship	
Electric Spark	1855	8	clipper	ship	Howes rig
Ellen Foster	1852	43	clipper	ship	
Elvira	1855	40	freighter	ship	
Emerald	1855	25	packet	ship	Howes rig on fore and main
Empress of the Sea	1853	39	clipper	ship	

Vessel	*Year*	*Spar Plan*	*Category*	*Rig*	*Remarks*
Endeavor	1856	43	clipper	ship	
Eringo	1853	65	clipper	bark	
Fair Wind	1855	8	clipper	ship	Howes rig
Fearless	1853	50	clipper	ship	
Flying Arrow	1852	34	clipper	ship	
Flying Childers	1852	9	clipper	ship	Forbes rig (original)
Flying Cloud	1851	31	clipper	ship	
Flying Dutchman	1852	31	clipper	ship	
Flying Eagle	1852	53	clipper	ship	
Flying Fish	1851	28	clipper	ship	
Flying Mist	1856	17	clipper	ship	Linnell's rig
Galatea	1854	53	clipper	ship	
Game Cock	1850	34	clipper	ship	
Gazelle	1851	46	clipper	ship	
Golden Fleece	1855	12	clipper	ship	Howes rig
Golden Light	1853	35	clipper	ship	
Good Hope	1855	23	freighter	ship	Howes rig on fore and main
Great Republic	1853	1	clipper	4-mast bark	Original. Forbes rig.
Great Republic	1855	2	clipper	4-mast bark	Rebuilt. Howes rig.
Halifax	1855	65	clipper	bark	
Harry Bluff	1855	53	packet	ship	
Henry Hill	1856	63	clipper	bark	
Herald of the Morning	1853	34	clipper	ship	
Hesperus	1856	18	freighter	ship	Howes rig
Hoogly	1851	53	clipper	ship	
Hurricane	1851	4	clipper	ship	
Invincible	1851	47	clipper	ship	
J. Montgomery	1852	49	packet	ship	
James Baines	1854	39	clipper	ship	
Japan	1854	49	packet	ship	
John Bertram	1850	37	clipper	ship	
John Gilpin	1852	49	clipper	ship	
John Land	1853	44	clipper	ship	
John Stuart	1851	53	clipper	ship	
Joseph Peabody	1856	8	clipper	ship	Howes rig
Joseph Walker	1850	45	packet	ship	
Josephine	1855	73	freighter	brigantine	
Kremlin	1850	64	freighter	bark	
Lady Suffolk	1855	49	packet	ship	
Lamplighter	1854	65	clipper packet	bark	
Lightning	1854	41	clipper	ship	
Mameluke	1855	8	clipper	ship	Howes rig
Manhasset	1852	75	freighter	schooner	
Mermaid	1851	58	clipper	bark	Forbes rig
Minna	1851	74	freighter	topsail schooner	
Moneynick	1854	66	packet	bark	
Mystery	1853	49	clipper	ship	
Nightingale	1851	32	clipper	ship	
Noonday	1855	43	clipper	ship	
Ocean Monarch	1856	19	freighter	ship	Howes rig
Orient	1852	49	packet	ship	
Phantom	1852	32	clipper	ship	
Plandome	1852	75	freighter	schooner	
Plutarch	1856	24	packet	ship	Lower topsail yards are stationary, slung with lifts.
Polar Star	1852	55	freighter	ship	

Vessel	*Year*	*Spar Plan*	*Category*	*Rig*	*Remarks*
Queen of Clippers	1853	31	clipper	ship	
Queen of the Seas	1852	49	clipper	ship	
Quickstep	1855	64	clipper	bark	
Racehorse	1850	58	clipper	bark	Forbes rig
Racer	1851	53	clipper	ship	
Radiant	1853	53	clipper	ship	
Red Jacket	1853	31	clipper	ship	
Roebuck	1851	53	clipper	ship	
Romance of the Sea	1853	38	clipper	ship	
Samuel Lawrence	1851	49	clipper	ship	
Saracen	1854	53	clipper	ship	
Sea Bird	1851	38	clipper	bark	
Sea Witch	1846	30	clipper	ship	
Seamans Bride	1851	3	clipper	ship	
Shooting Star	1851	47	clipper	ship	
Sierra Nevada	1854	22	clipper	ship	Howes rig
Silver Star	1856	31	clipper	ship	
Southern Cross	1851	32	clipper	ship	
Sovereign of the Seas	1852	39	clipper	ship	
Spitfire	1853	37	clipper	ship	
Staffordshire	1851	49	clipper packet	ship	
Stag Hound	1850	27	clipper	ship	
Star of Empire	1853	53	packet	ship	
Storm King	1853	49	clipper	ship	
Sultana	1850	60	clipper	bark	
Surprise	1850	27	clipper	ship	
Susan Howland	1856	49	freighter	ship	
Sweepstakes	1853	26	clipper	ship	
Sword Fish	1851	34	clipper ship		
		14	clipper	ship	Howes rig, fore and main only. Apparently a re-rig, undated.
Syren	1851	34	clipper	ship	
Thatcher Magoun	1856	8	clipper	ship	Howes rig
Trieste	1857	61	freighter	bark	
Union	1850	48	freighter	ship	
Vitula	1855	13	freighter	ship	Howes rig
Volante	1852	71	freighter	brig	
War Hawk	1855	42	clipper	ship	
Water Witch	1853	29	clipper	ship	
Webster	1853	49	packet	ship	
West Wind	1853	44	clipper	ship	
Westward Ho	1852	31	clipper	ship	
Wild Pigeon	1851	28	clipper	ship	
Wildfire	1853	62	clipper	bark	
Winchester	1851	52	freighter	ship	
Winfield Scott	1851	54	packet	ship	
Winged Arrow	1852	44	clipper	ship	
Winged Racer	1852	34	clipper	ship	
Witch of the Wave	1851	28	clipper	ship	
Witch of the Wave	1856	10	freighter	ship	Howes rig
Witchcraft	1850	29	clipper	ship	
Young America	1853	5	clipper	ship	
Young Turk	1856	57	clipper	bark	Humphrey's rig

Spar Plan #3—Three masts, ship; single topsails; two jibs; moonsails on fore, main and mizzen.

Applies to *Seaman's Bride*, 1851.

Spar Plan #4—Three masts, ship; single topsails; two jibs; moonsails on fore and main; skysail on mizzen.

Applies to the clipper *Hurricane*, 1851.

Spar Plan #5—Three masts, ship; single topsails; two jibs; moonsail on main; skysails on fore and mizzen. Topgallant masts are headed (fitted with crosstrees); main spencer; monkey gaff.

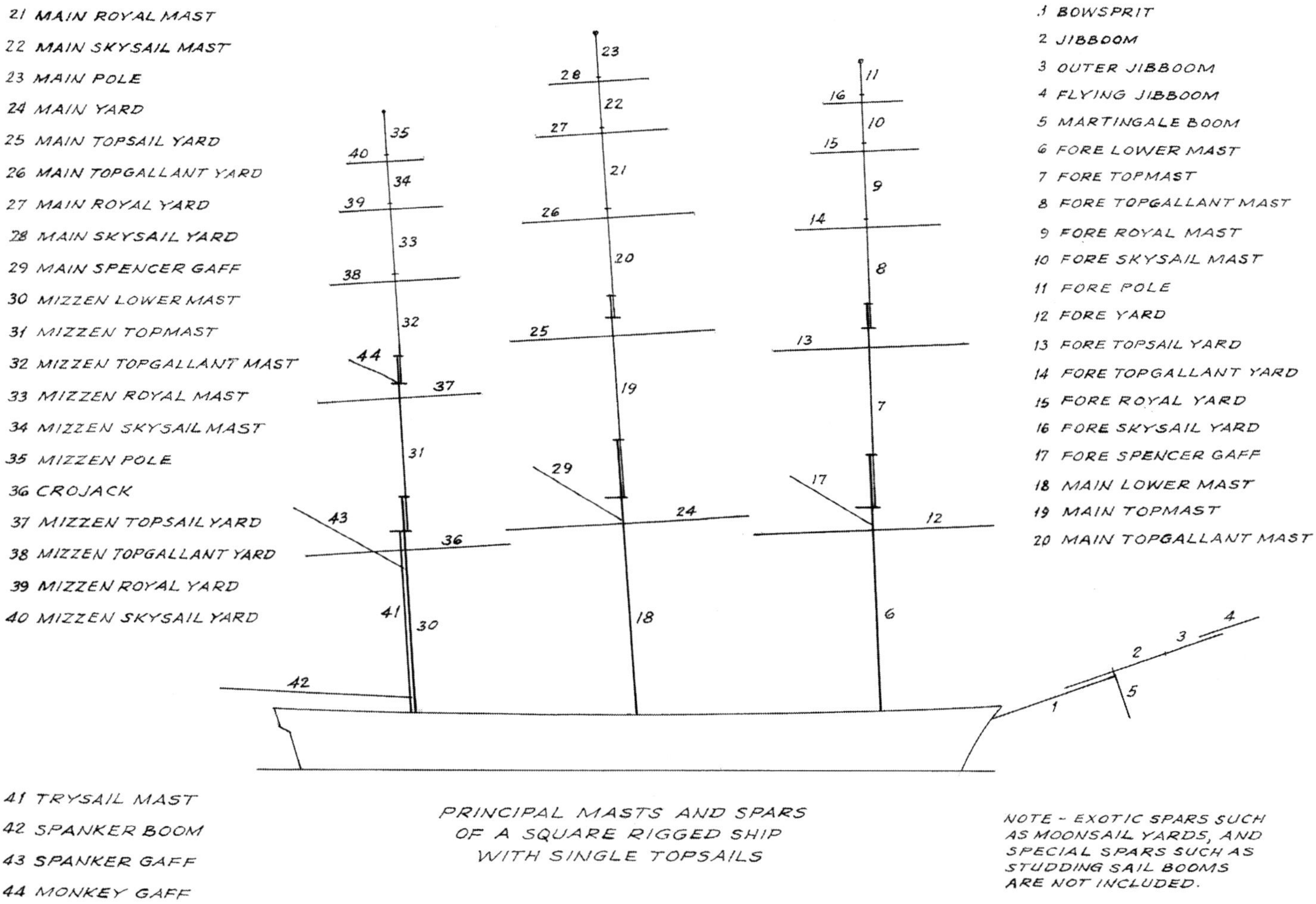

Figure 11.1. Principal masts and spars of a square rigged ship with single topsails.

Applies to the clipper *Young America*, 1853.

Spar Plan #6—Three masts, ship; Forbes rig, with upper masts fidded abaft the lower masts; two jibs; skysails on fore, main and mizzen.

Applies to the clipper *Belle of the West*, 1853. All references are not in agreement. See Spar Plan #7.

Spar Plan #7—Three masts, ship; Forbes rig; two jibs; skysails on fore, main and mizzen.

Applies to the clippers *Belle of the West*, 1853 (all references are not in agreement, See Spar Plan #6), and *Amphitrite*, 1853.

Spar Plan #8—Three masts, ship; Howes rig; two jibs; skysails on fore, main and mizzen.

Applies to the clippers *Asa Eldridge*, 1856; *Electric Spark*, 1855; *Joseph Peabody*, 1856; *Fair Wind*, 1855; *Thatcher Magoun*, 1856; *Alarm*, 1856, *Mameluke*, 1855

Spar Plan #9—Three masts, ship; Forbes rig, with fore and main upper masts fidded abaft the lower masts, and mizzen upper mast fidded afore the lower mast; two jibs; skysails on fore and main; royal on the mizzen; trysail mast.

Applies to the clipper *Flying Childers*, 1852

Spar Plan #10—Three masts, ship; Howes rig; two jibs; skysail on main; royals on fore and mizzen; main spencer.

Applies to the freighter *Witch of the Wave*, 1856.

Spar Plan #11—Three masts, ship; Howes rig; three jibs in a single spar; skysail on main; royals on fore and mizzen.

Applies to the clipper packet *Donald McKay*, 1855.

Spar Plan #12—Three masts, ship; Howes rig; two jibs; skysail on main; royals on fore and mizzen.

Applies to the clippers *Golden Fleece*, 1855, and *Andrew Jackson*, 1855.

Spar Plan #13—Three masts, ship; Howes rig; two jibs; skysails fore, main and mizzen, rigged aloft.

Applies to the clipper *Climax*, 1853, and the freighter *Vitula*, 1855.

Spar Plan #14—Three masts, ship; Howes rig on fore and main; single topsail on mizzen; two jibs; skysails on fore, main and mizzen.

Applies to the clipper *Sword Fish*, 1851, and was apparently a rerig, date unknown.

Spar Plan #15—Three masts, ship; Howes rig on fore and main; single topsail on mizzen; two jibs; skysail on main; royals on fore and mizzen.

Applies to the clipper *Blue Jacket*, 1854.

Spar Plan #16—Three masts, ship; Forbes rig, original, upper masts fidded abaft the lower masts; two jibs; royals on fore, main and mizzen.

Applies to the packet *Cornelius Grinnell*, 1850.

Spar Plan #17—Three masts, ship; Linnell's rig (topsail yards resemble Howes rig); two jibs; skysail on main only, rigged aloft.; royals on fore and mizzen.

Applies to the clipper *Flying Mist*, 1856.

Spar Plan #18—Three masts, ship; Howes rig; two jibs; skysail on main, rigged aloft.; royals on fore and mizzen.

Applies to the freighter *Hesperus*, 1856.

Spar Plan #19—Three masts, ship; Howes rig; three jibs in a single spar; royals on fore, main and mizzen; trysail mast; main spencer.

Applies to the freighter *Ocean Monarch*, 1856.

Spar Plan #20—Three masts, ship; Howes rig; two jibs; sliding gunter royals on fore, main and mizzen.

Applies to the packet *Commodore Perry*, 1854.

Spar Plan #21—Three masts, ship; Forbes rig; two jibs; royals on fore, main and mizzen.

Applies to the packet *Aurora*, 1854.

Spar Plan #22—Three masts, ship; Howes rig; two jibs; royals on fore, main and mizzen.

Applies to the clipper *Sierra Nevada*, 1854.

Spar Plan #23—Three masts, ship; Howes rig on fore and main; single topsail on mizzen; two jibs; royals on fore, main and mizzen; main spencer.

Applies to the freighter *Good Hope*, 1855.

Spar Plan #24—Three masts, ship; two jibs; double topsails, lower topsail yards are stationary, slung with lifts; royals on fore, main and mizzen.

Applies to the packet *Plutarch*, 1856, and the freighter *Dione*, 1857.

Spar Plan #25—Three masts, ship; Howes rig on fore and main; single topsail on mizzen; two jibs; royals on fore, main and mizzen.

Applies to the packet *Emerald*, 1855.

Spar Plan #26—Three masts, ship; single topsails; three jibs, flying jibboom a separate spar; skysails on fore, main and mizzen; topgallant masts are headed (fitted with crosstrees).

Applies to the clipper *Sweepstakes*, 1853.

Spar Plan #27—Three masts, ship; single topsails; three jibs, flying jibboom a separate spar; skysails on fore, main and mizzen; fore and main spencers.

Applies to the clippers *Stag Hound*, 1850 and *Surprise*, 1850.

Spar Plan #28—Three masts, ship; single topsails; two jibs; skysails on fore, main and mizzen; fore and main spencers.

Applies to the clippers *Comet*, *Flying Fish*, *Wild Pigeon* and *Witch of the Wave*, all built in 1851.

Spar Plan #29—Three masts, ship; single topsails; three jibs, flying jibboom a separate spar; skysails on fore, main and mizzen; main spencer.

Applies to the clippers *Witchcraft*, 1850; *Challenge*, 1851; and *Water Witch*, 1853.

Spar Plan #30—Three masts, ship; single topsails; three jibs in a single spar; skysails on fore main and mizzen; main spencer.

Applies to the clipper *Sea Witch*, 1846.

Spar Plan #31—Three masts, ship; single topsails; two jibs; skysails on fore, main and mizzen; main spencer.

Applies to the clippers *Flying Cloud*, 1851; *Westward Ho*, 1852; *Flying Dutchman*, 1852; *Red Jacket*, 1853; *Queen of Clippers*, 1853; *Silver Star*, 1856

Spar Plan #32—Three masts, ship; single topsails; three jibs, flying jibboom a separate spar; skysails on fore, main and mizzen.

Applies to the clippers *Nightingale*, 1851; *Southern Cross*, 1851; and *Phantom*, 1852.

Spar Plan #33—Three masts, ship; single topsails; three jibs in a single spar; skysails on fore, main and mizzen.

Applies to the clipper *Coeur de Lion*, 1854.

Spar Plan #34—Three masts, ship; single topsails; two jibs; skysails on fore, main and mizzen.

Applies to the clippers *Herald of the Morning*, 1853; *Game Cock*, 1850; *Syren*, 1851; *Winged Racer*, 1852; *Sword Fish*, 1851; *Challenger*, 1853; *Dauntless*, 1852; *Edwin Forrest*, 1853; *Flying Arrow*, 1852

Spar Plan #35—Three masts, ship; single topsails; two jibs; skysails on fore, main and mizzen, rigged aloft.; main spencer.

Applies to the clipper *Golden Light*, 1853.

Spar Plan #36—Three masts, ship; single topsails; three jibs, flying jibboom a separate spar; skysails on fore, main and mizzen, rigged aloft.

Applies to the clipper *Charmer*, 1854.

Spar Plan #37—Three masts, ship; single topsails; two jibs; skysails on fore, main and mizzen, rigged aloft.

Applies to the clippers *John Bertram*, 1850, and *Spitfire*, 1853.

Spar Plan #38—Three masts, ship; single topsails; two jibs; skysails on main; royals on fore and mizzen; fore and main spencers.

Applies to the clipper *Romance of the Sea*, 1853.

Spar Plan #39—Three masts, ship; single topsails; two jibs; skysail on main; royals on fore and mizzen; main spencer.

Applies to the clippers *Sovereign of the Seas*, 1852; *Empress of the Sea*, 1853; and *James Baines*, 1854.

Spar Plan #40—Three masts, ship; single topsails; two jibs; skysail on main; royals on fore and mizzen; fore spencer.

Applies to the freighter *Elvira*, 1855.

Spar Plan #41—Three masts, ship; single topsails; three jibs, flying jibboom a separate spar; skysail on main; royals on fore and mizzen.

Applies to the clipper *Lightning*, 1854.

Spar Plan #42—Three masts, ship; single topsails; three jibs in a single spar; skysail on main; royals on fore and mizzen.

Applies to the clipper *War Hawk*, 1855.

Spar Plan #43—3 masts, ship; single topsails; two jibs; skysail on main; royals on fore and mizzen.

Applies to the clippers *Bald Eagle*, 1852; *Ellen Foster*, 1852; *Eagle Wing*, 1853; *Noonday*, 1855; *Endeavor*, 1856.

Spar Plan #44—Three masts, ship; single topsails; two jibs; skysail on main, rigged aloft.; royals on fore and mizzen.

Applies to the clippers *Winged Arrow*, 1852; *John Land*, 1853 and *West Wind*, 1853.

Spar Plan #45—Three masts, ship; single topsails; two jibs; royals on fore, main and mizzen; trysail mast; main spencer.

Applies to the packet *Joseph Walker*, 1850.

Spar Plan #46—Three masts, ship; single topsails; two jibs; royals on fore, main and mizzen; monkey gaff; main spencer.

Applies to the clippers *Gazelle*, 1851, and *Black Hawk*, 1856.

Spar Plan #47—Three masts, ship; single topsails; three jibs, flying jibboom a separate spar; royals on fore, main and mizzen; main spencer.

Applies to the clippers *Invincible*, 1851, and *Shooting Star*, 1851.

Spar Plan #48—3 masts, ship; single topsails; two jibs; flying jibboom a separate spar; royals on fore, main and mizzen; main spencer.

Applies to the freighter *Union*, 1850.

Spar Plan #49—Three masts, ship; single topsails; two jibs; royals on fore, main and mizzen; main spencer.

Applies to the following vessels:

Clippers	*Packets*
Samuel Lawrence, 1851	*Constantine*, 1850
John Gilpin, 1852	*Clara Wheeler*, 1850
J. *Montgomery*, 1852	*Orient*, 1852
Queen of the Seas, 1852	*Japan*, 1854
Mystery, 1853	*Webster*, 1853
Storm King, 1853	*Lady Suffolk*, 1855
Champion of the Seas, 1854	
Clipper Packet	*Freighter*
Staffordshire, 1851	*Susan Howland*, 1856

Spar Plan #50—Three masts, ship; single topsails; three jibs in a single spar; royals on fore, main and mizzen; monkey gaff.

Applies to the clipper *Fearless*, 1853.

Spar Plan #51—Three masts, ship; single topsails; three jibs in a single spar; royals on fore, main and mizzen.

Applies to the freighter *Alma*, 1855.

Spar Plan #52—Three masts, ship; single topsails; three jibs, flying jibboom a separate spar; royals on fore, main and mizzen.

Applies to the freighter *Winchester*, 1851.

Spar Plan #53—Three masts, ship; single topsails; two jibs; royals on fore, main and mizzen.

Applies to the following vessels:

Clippers	*Packets*
Hoogly, 1851	*Star of the Empire*, 1853
John Stuart, 1851	*Barreda Brothers*, 1854
Racer, 1851	*Chatsworth*, 1854
Flying Eagle, 1853	*Abbott Lawrence*, 1855
Galatea, 1853	*Amos Lawrence*, 1855
Radiant, 1853	*Cathedral*, 1855
Roebuck, 1853	*Harry Bluff*, 1855
Saracen, 1854	

Spar Plan #54—Three masts, ship; single topsails; three jibs; royals on fore, main and mizzen, rigged aloft.

Applies to the packet *Winfield Scott*, 1851.

Spar Plan #55—Three masts, ship; single topsails; two jibs, flying jib a separate spar; royal on the main; topgallants on fore and mizzen.

Applies to the freighter *Polar Star*, 1852.

Spar Plan #56—Three masts, ship; single topsails; two jibs; topgallant yards on fore, main and mizzen.

Applies to the packet *Antarctic*, 1850.

Spar Plan #57—Three masts, bark; Humphrey's rig; double topsails on fore and main; two jibs; royals on fore and main; mizzen topmast is a pole.

Note: Humphrey's rig featured additional stub topmasts below the fore and main caps. See Fig. 6.9.

Applies to the clipper bark *Young Turk*, 1856.

Spar Plan #58—Three masts, bark; Forbes rig; double topsails, fore and main; two jibs; royals on fore and main; mizzen topmast is a pole; main spencer. See Fig. 6.9.

Applies to the clipper barks *Racehorse*, 1850, and *Mermaid*, 1851.

Spar Plan #59—Three masts, bark; Howes rig; double topsails on fore and main; two jibs; royals on fore and main; mizzen topmast is a pole. See Fig. 6.9.

Applies to the freighter *Aurelia*, 1855.

Spar Plan #60—Three masts, bark; single topsails; three jibs; flying jibboom is a separate spar; skysail on main; royal on fore; main spencer; mizzen topmast is a pole.

Applies to the clipper bark *Sultana*, 1850.

Spar Plan #61—Three masts, bark; single topsails; two jibs; skysail on main; royal on fore; fore spencer; mizzen topmast is a pole.

Applies to the freighter *Trieste*, 1857.

Spar Plan #62—Three masts, bark; single topsails; two jibs; skysails on fore and main, rigged aloft.; mizzen topmast is a pole.

Applies to the clipper bark *Wildfire*, 1853.

Spar Plan #63—Three masts, bark; single topsails; two jibs; main skysail, rigged aloft.; royal on foremast; mizzen topmast is a pole.

Applies to the clipper bark *Henry Hill*, 1856.

Spar Plan #64—Three masts, bark; single topsails; two jibs; royals on fore and main; main spencer; mizzen topmast is a pole.

Applies to the freighter *Kremlin*, 1850, and the clipper bark *Quickstep*, 1855.

Spar Plan #65—Three masts, bark; single topsails; two jibs; royals on fore and main; mizzen topmast is a pole.

Applies to the clipper barks *Eringo*, 1853, *Bounding Billow*, 1854, and *Halifax*, 1855; the clipper packet *Lamplighter*, 1854; and the freighters *Benin*, 1854, and *Alice Tainter*, 1856.

Spar Plan #66—Three masts, bark; single topsails; two jibs; royals on fore and main, rigged aloft.; mizzen topmast is a pole.

Applies to the packet *Moneynick*, 1854.

Spar Plan #67—Three masts, bark; single topsails; two jibs; royal on main; topgallant on fore; main spencer; mizzen topmast is a pole.

Applies to the packet *Edisto*, 1851.

Spar Plan #68—Three masts, bark; single topsails; two jibs; topgallants on fore and main; mizzen topmast is a pole.

Applies to the clipper bark *Sea Bird*, 1851.

Spar Plan #69—Three masts, "barkentine." (At this period the term barkentine had not yet come into use. The vessel was referred to originally as a "demi-barque." See the Chronological List.); Forbes rig on foremast; two jibs; royal on the foremast; boom foresail; gaff mainsail; spanker; main and mizzen topmasts are poles.

Applies to the freighter *Antelope*, 1855.

Spar Plan #70—Two masts, brig, fitted with a trysail mast (possibly more correctly referred to as a snow); single topsails; two jibs; royals on fore and main.

Applies to the packet *Belle*, 1850.

Spar Plan #71—Two masts, brig; two jibs, flying jibboom a separate spar; royals on fore and main.

Applies to the freighter *Volante* 1852.

Spar Plan #72—Two masts, brig; single topsails; two jibs; royals on fore and main, rigged aloft.

Applies to the clipper brig *Belle*, 1853. The vessel is described as having her masts stepped with "Baltimore clipper rake."

Spar Plan #73—Two masts, brigantine; single topsail; two jibs; royal on foremast; boom mainsail; main topmast is a pole.

Applies to the freighter *Josephine*, 1855.

Spar Plan #74—Two masts, topsail schooner; two jibs; square topsail on foremast; boom foresail and mainsail; main topmast is a pole.

Applies to the freighters *Brenda*, 1851, and *Minna*, 1851, both of which were built from the same plans.

Spar Plan #75—Two masts, schooner; two jibs; boom mainsail; gaff foresail; fore and main topmasts are poles.

Applies to the freighters *Plandome*, 1851, and *Manhasset*, 1851, both of which were built from the same plans.

Of all the sparring arrangements shown three variations can be described as unusual. These are Forbes double topsail masting with the upper masts fidded abaft the mast immediately below, as shown in arrangements #6, #9 and #16; double topsail masting with the lower topsail yard slung in permanent location as shown in arrangement #24; and Humphrey's double topsail masting with the addition of stub topmasts, as shown in arrangement #57.

In 1844, Forbes introduced his double topsail rig, which involved lengthening of the lower mastheads, fitting an additional topsail yard, and fidding the upper masts abaft each mast below. Fidding the masts in this manner allowed for convenience in housing them.

This new rig immediately found favor and, by 1850, was in common usage, with one modification, that being the manner of fidding the masts. Even though housing the masts was improved, the aesthetic appearance of the masts fidding abaft did not find universal acceptance. As a result most ships adopted the concept of double topsails but rejected this new mode of installing masts.

Forbes remained an avid supporter of his original method, writing and publishing many articles on the subject, never yielding in his belief that fidding masts abaft was the best procedure.

His double topsail rig went unchallenged until 1853 when Capt. Frederic Howes introduced, in the clipper *Climax*, his own method of rigging double topsails. See Fig. 6.8.

Howes' method met with immediate acceptance and, in a short time, became the standard form of double topsail construction.

Some years later Forbes wrote that "the verdict of the public (generally, if not always governed by dollars and cents) has proved the Howes rig to be the most acceptable."

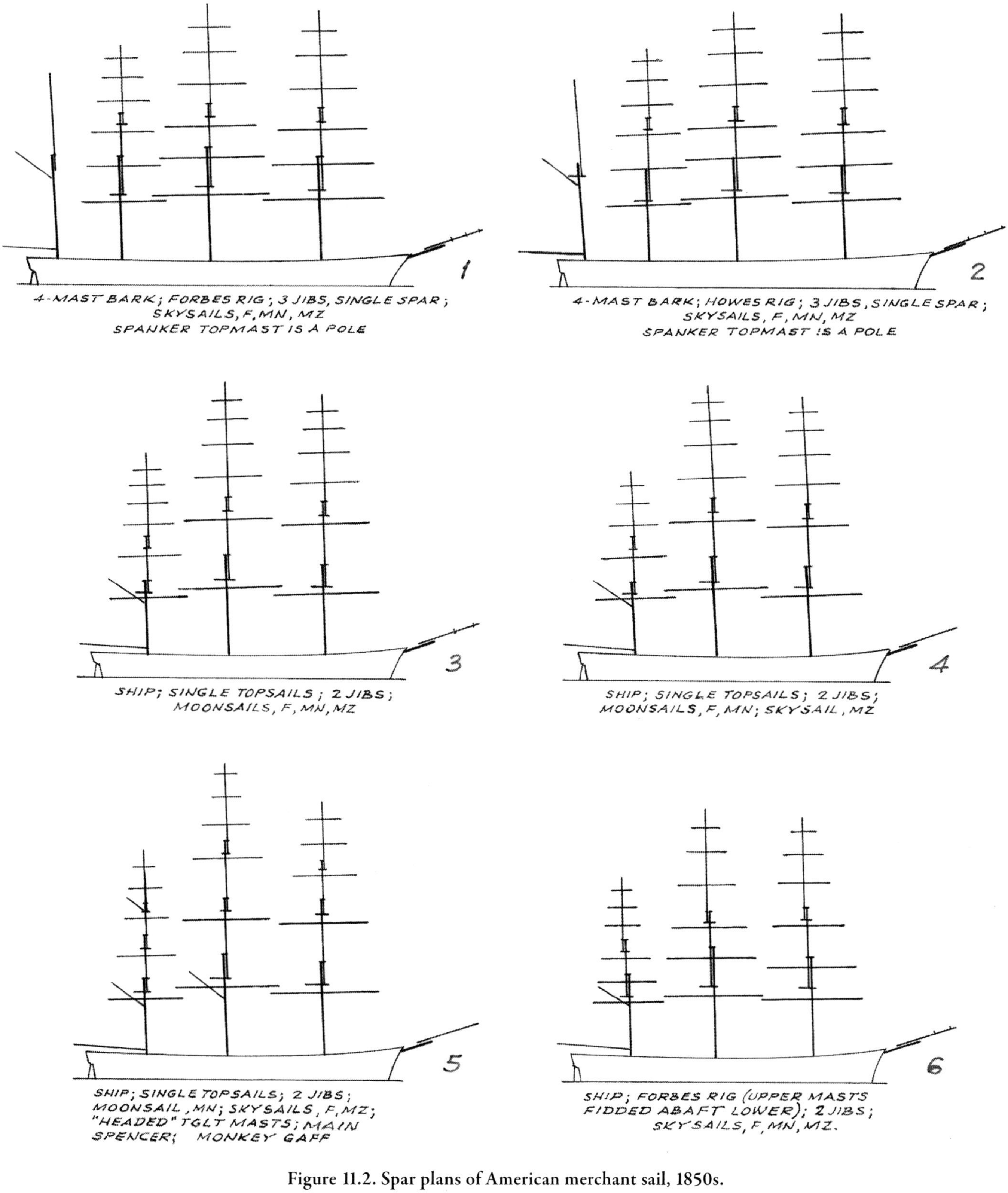

Figure 11.2. Spar plans of American merchant sail, 1850s.

Three examples of Forbes' original masting appear in this book, they being plans #6, #9 and #16.

Another unusual sparring arrangement, shown in plan #24, appeared in 1856 and again in 1857 in two separate vessels. These were the ships *Plutarch*, 1856, and *Dione*, 1857. Both vessels were built in the yard of G. and T. Boole, East Boston, *Plutarch* for Howland and Ridgeway, of New York, and *Dione* for Capt. Fales, of Boston.

The peculiarity in these ships, which were both rigged with double topsails, is the manner in which the lower topsail

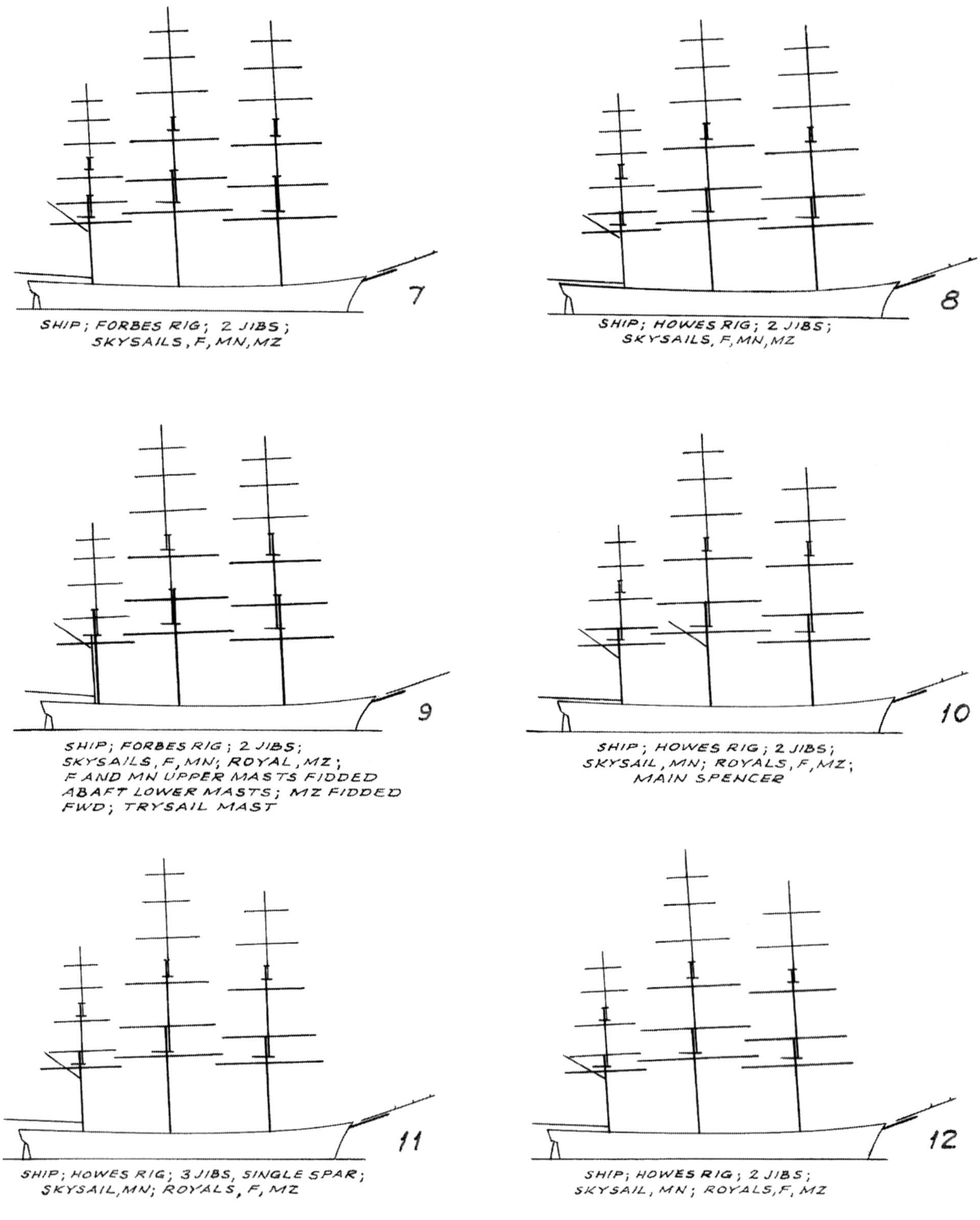

Figure 11.3. Spar plans of American merchant sail, 1850s.

yards were slung. In each case these yards were intended to be stationary, i.e., they would function at a fixed height, in the manner of the Howes rig, but they were not rigidly secured to the mast at that height.

Following are the descriptions of each installation. *Plutarch* "has double topsail yards, the lower ones slung to the topmasts instead of the caps, like Howes rig, though the principle is the same, as the yards are stationary."[1] *Dione*'s, "lower topsail yards are above the caps, and are hung by chains to the topmasts, and the upper yards hoist and lower in the usual manner."[1]

It can reasonably be assumed that the yard required a tye

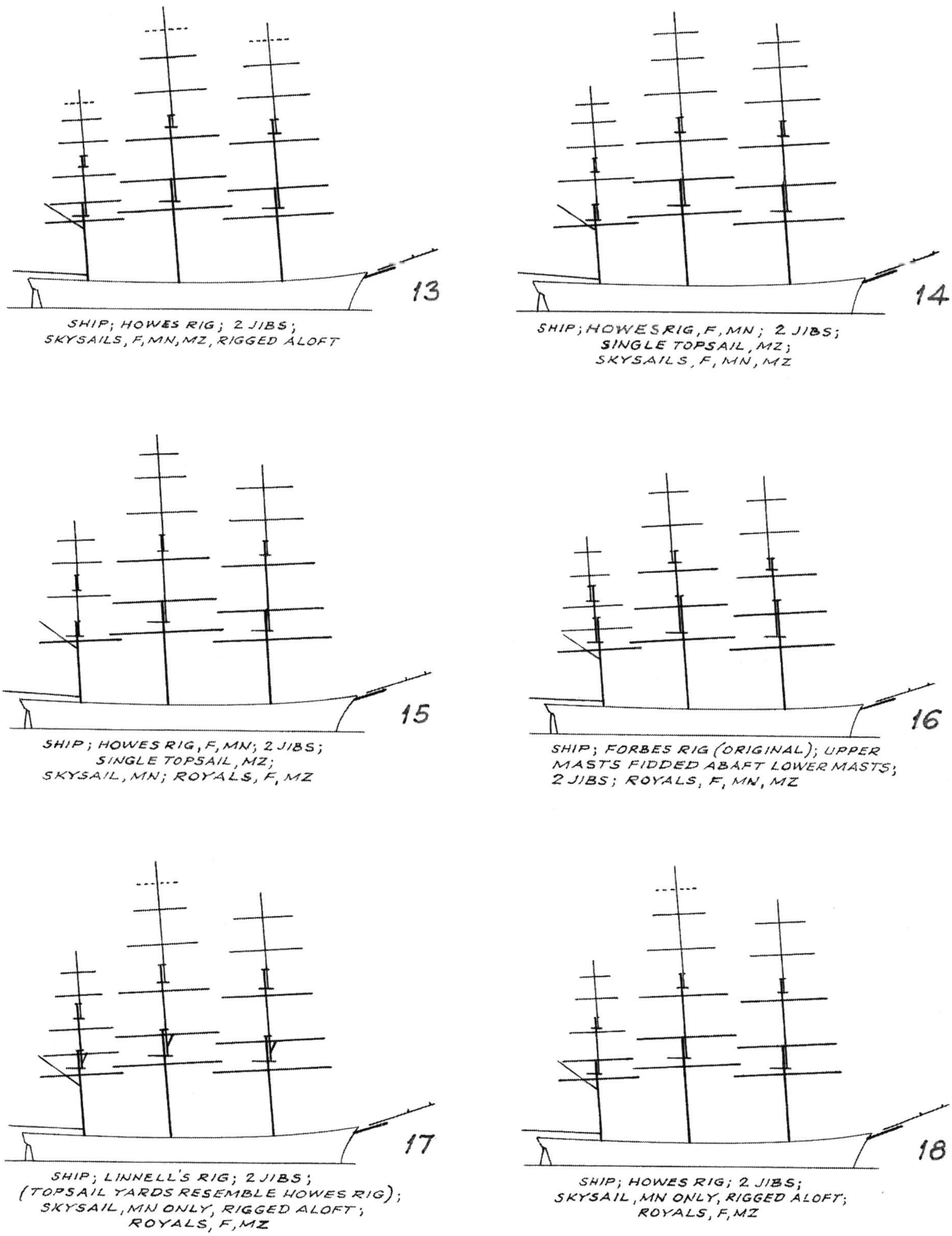

Figure 11.4. Spar plans of American merchant sail, 1850s.

to hoist it into position and a parral to confine it to the mast, in addition to the usual running rigging, such as braces and lifts.

It is not known who is responsible for this particular feature. It was never given a specific identity or claimed as the invention of an individual.

A close inspection of spar plans #2, #6 through #23, and #25, all of which feature double topsails, reveals that none of these shows lower topsail yards slung from the topmasts between the lower cap and the topmast crosstrees. Only spar plan #24 is fitted in this manner. This is a feature that could be overlooked by all but the practiced eye.

Finally, a very unusual example of masting is shown in spar plan #57. This involves the addition of a stub topmast. See Fig. 6.8, Humphrey's rig.

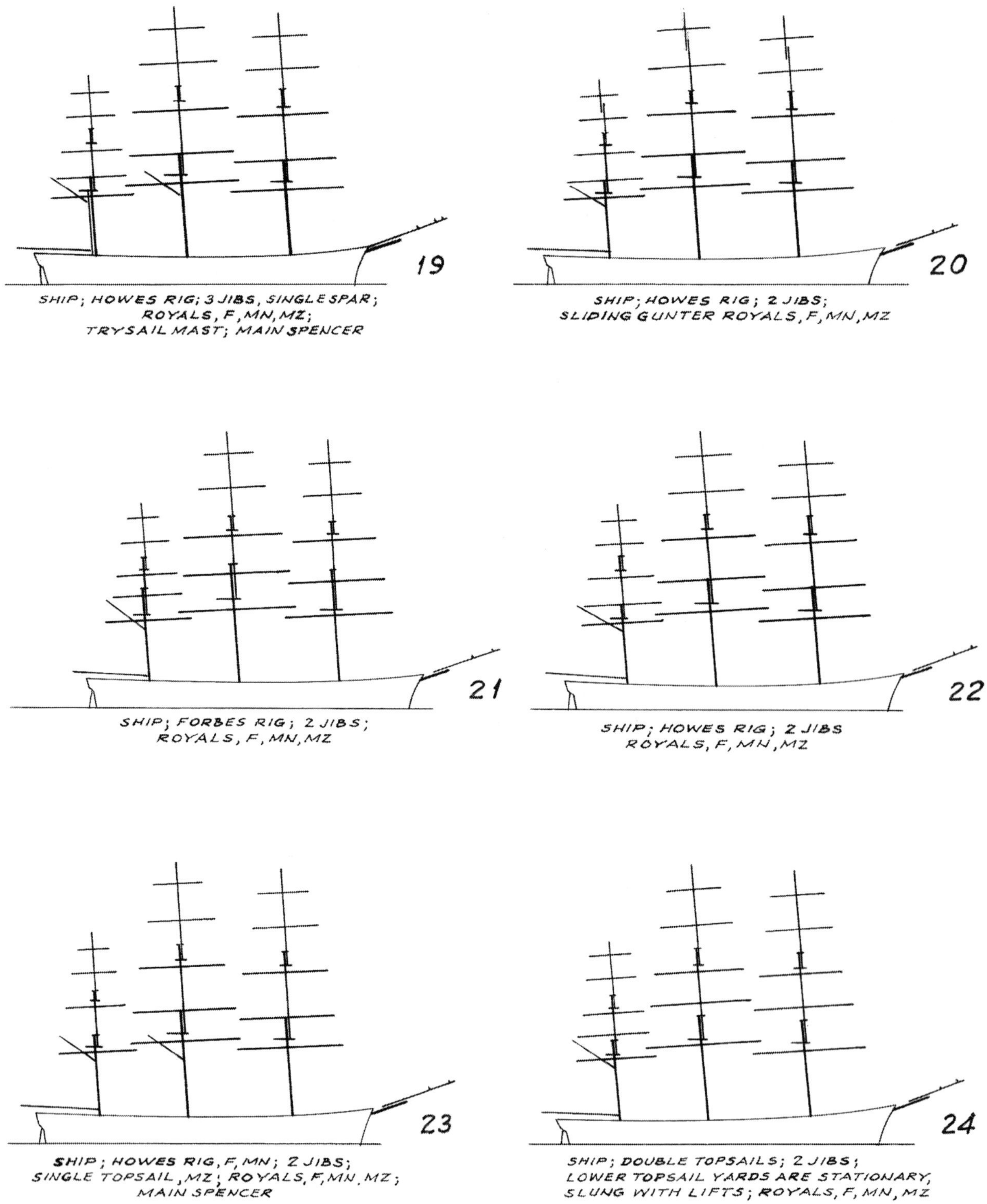

Figure 11.5. Spar plans of American merchant sail, 1850s.

Capt. John Humphrey, of Boston, invented and patented his own version of the double topsail rig.

A description of it[1] consists of only a few lines: "The new rig consists of double topmasts and double topsail yards. The lower topmast is just long enough to set the lower topsail."

This version of the double topsail was introduced in November 1856 in the small clipper bark *Young Turk* and is the only example uncovered to date.

This concludes the portion of this book that is devoted to the external shape, and the complex inner construction, of wooden masts and spars of the 1850s.

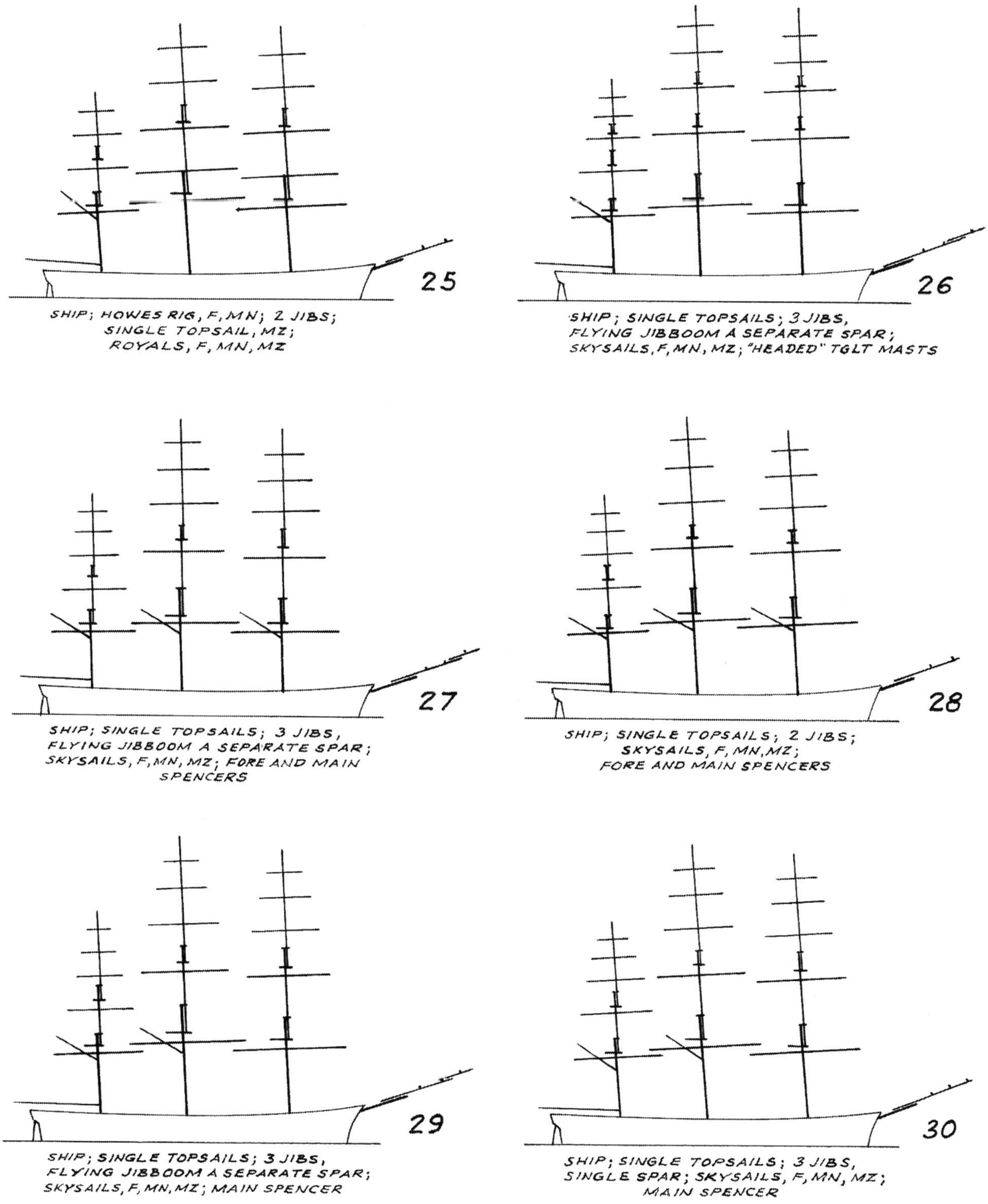

Figure 11.6. Spar plans of American merchant sail, 1850s.

The following pages are devoted to mast makers (Chapter 12) and masting lists, in the Appendix, which are relevant to the subject.

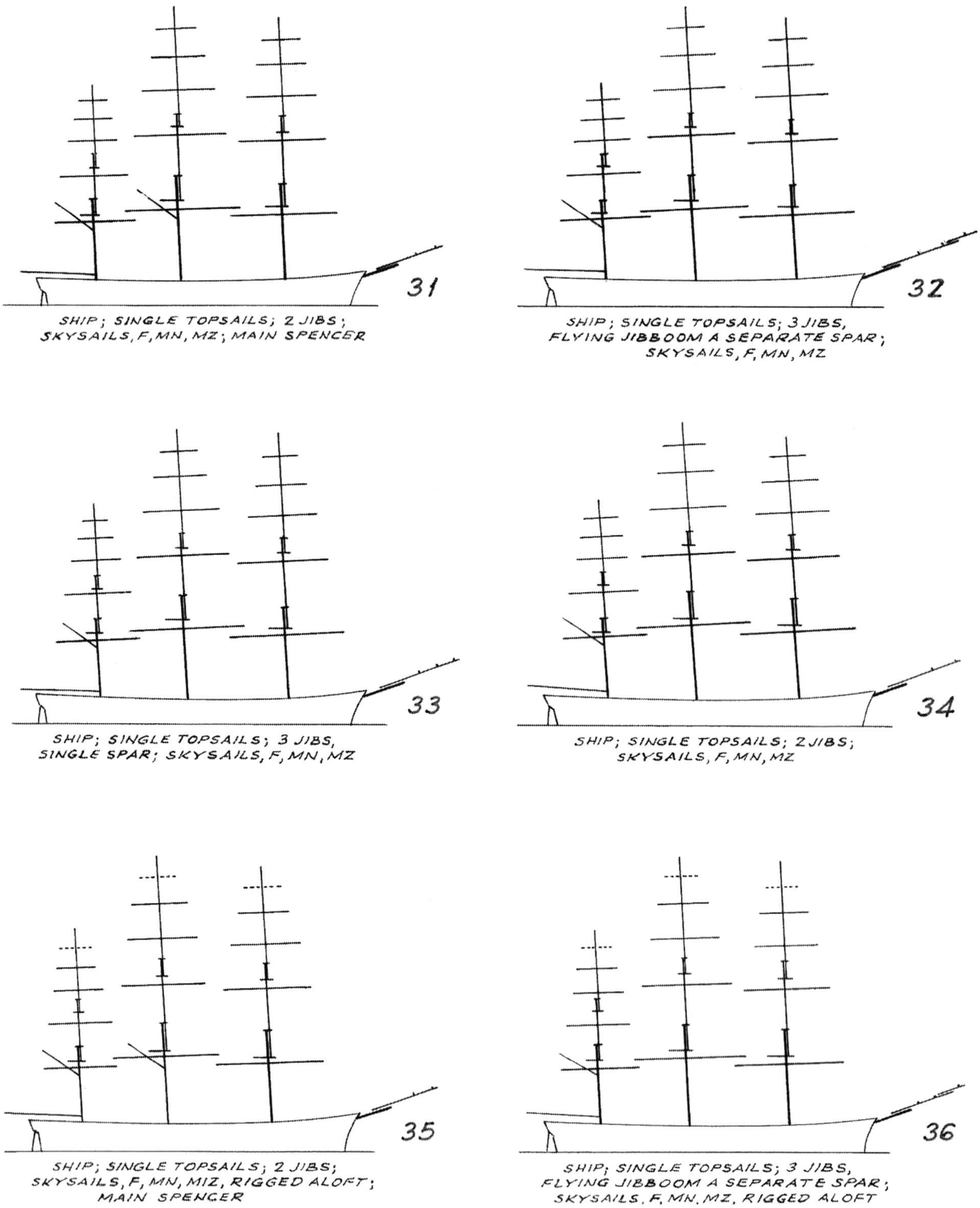

Figure 11.7. Spar plans of American merchant sail, 1850s.

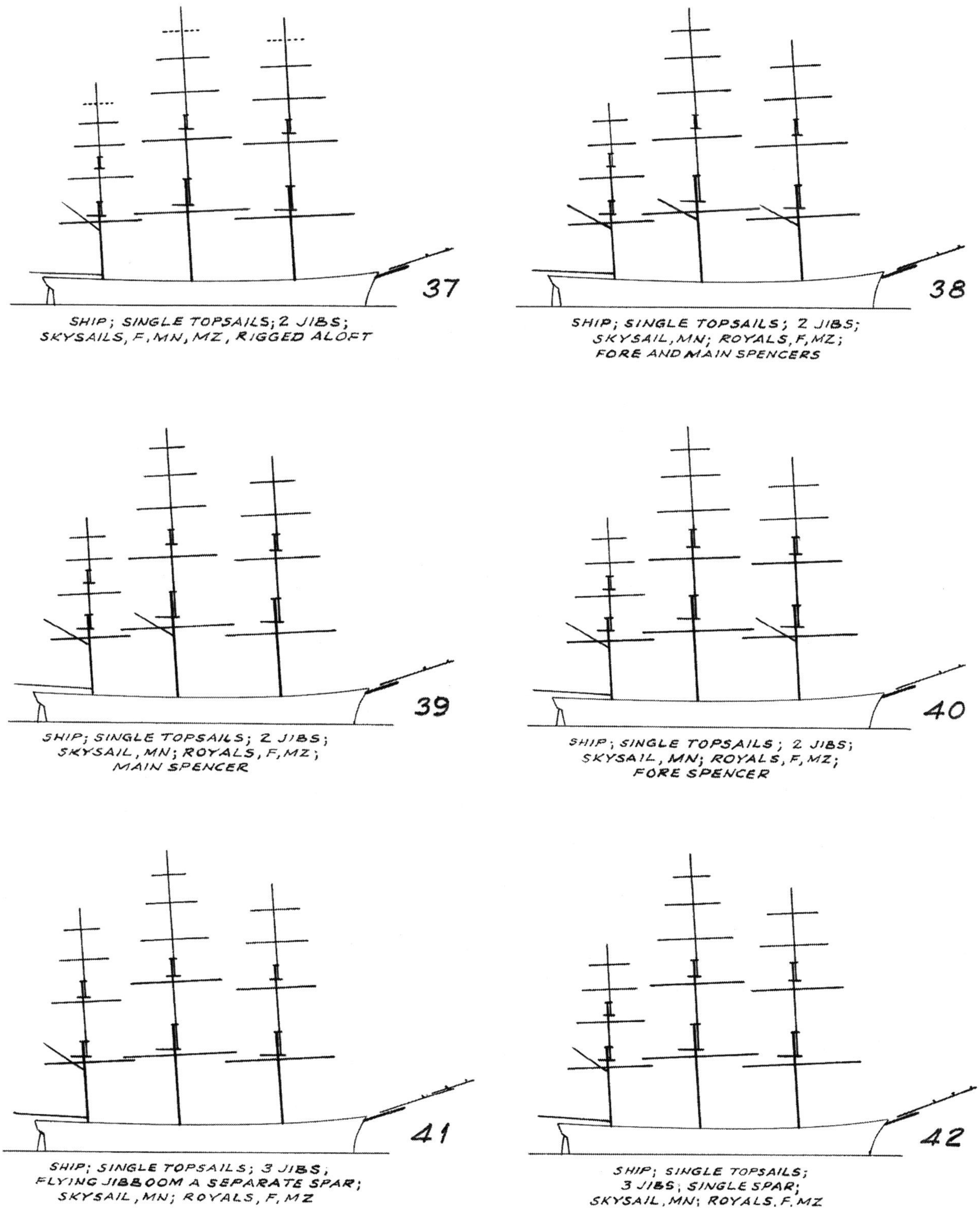

Figure 11.8. Spar plans of American merchant sail, 1850s.

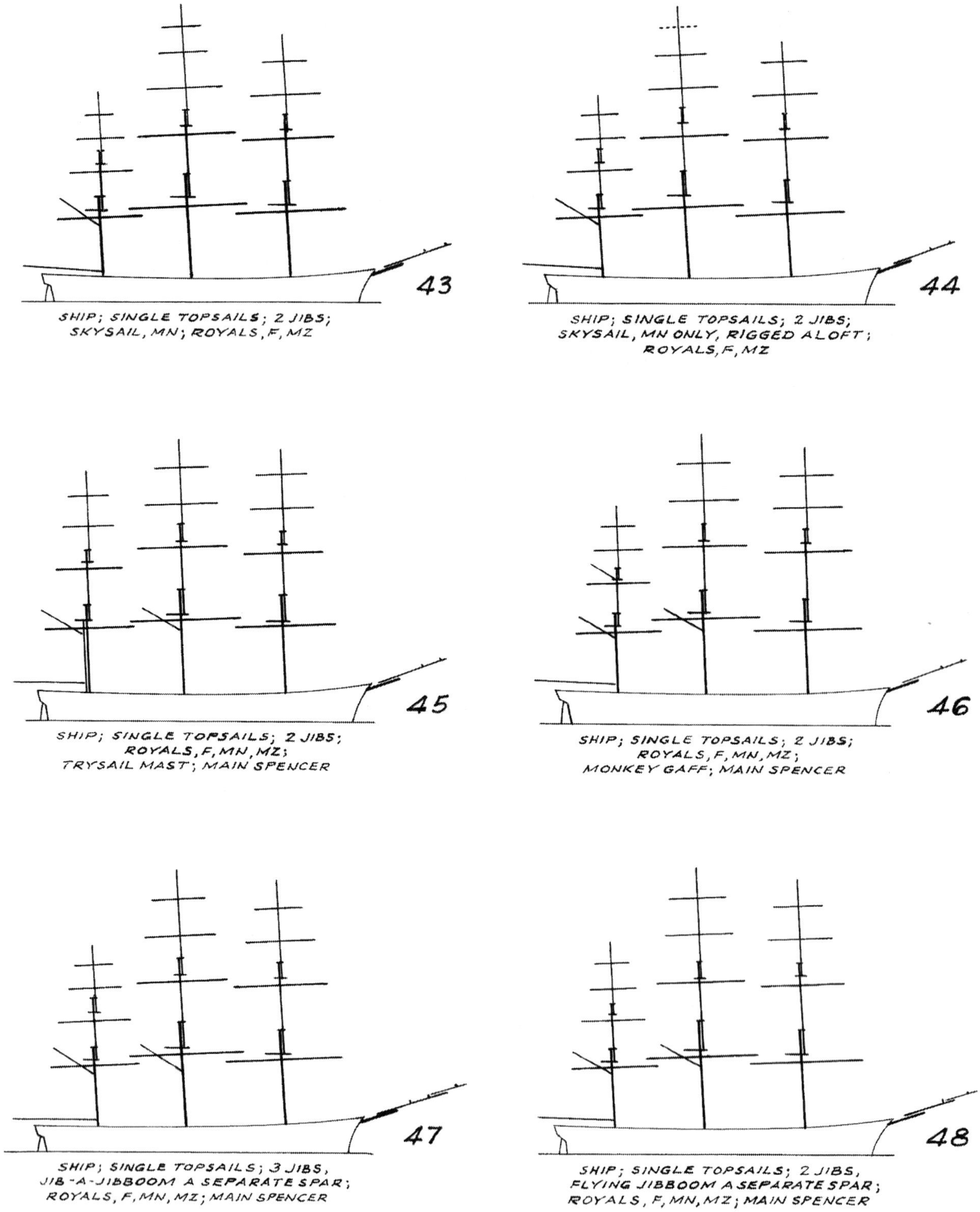

Figure 11.9. Spar plans of American merchant sail, 1850s.

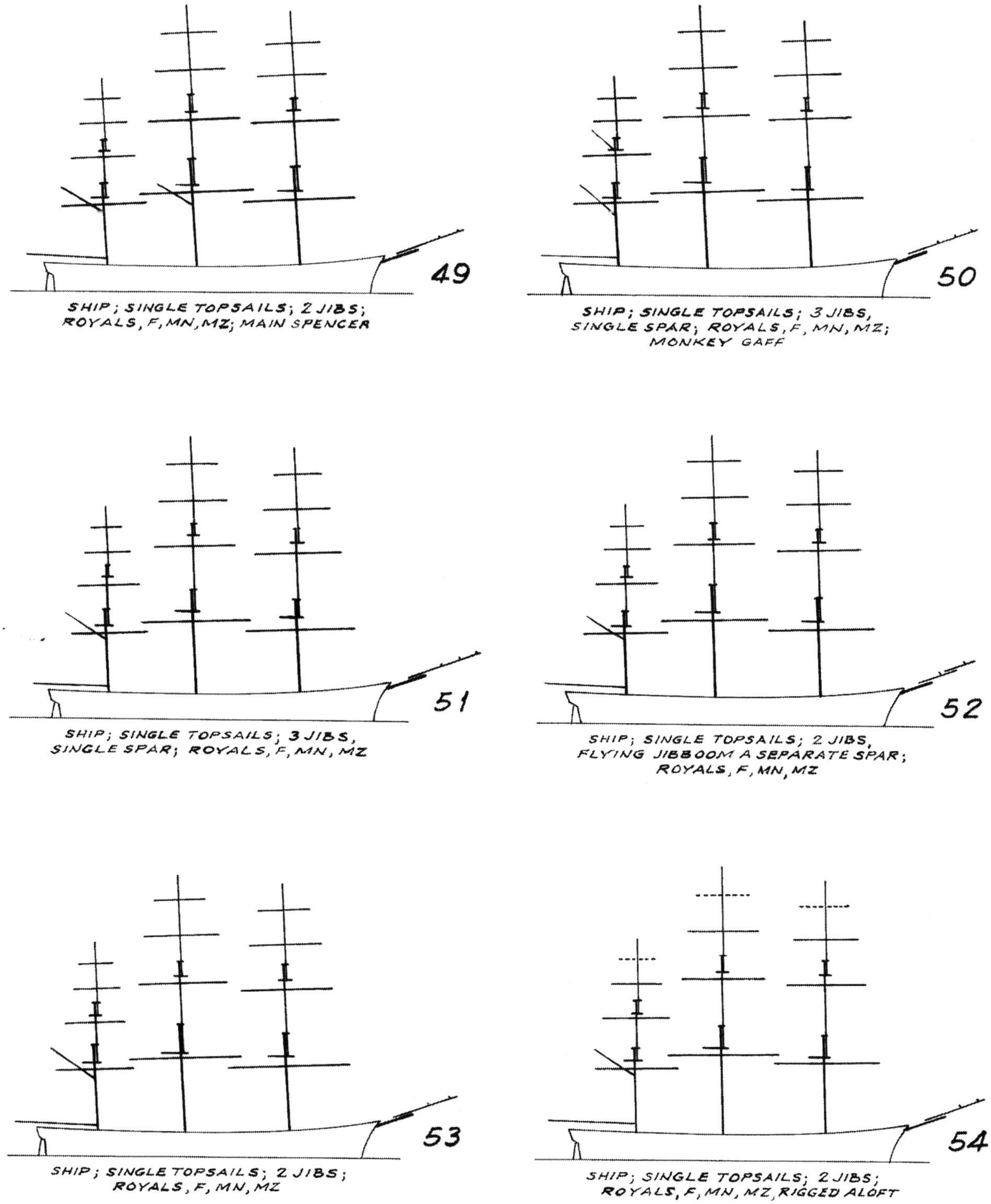

Figure 11.10. Spar plans of American merchant sail, 1850s.

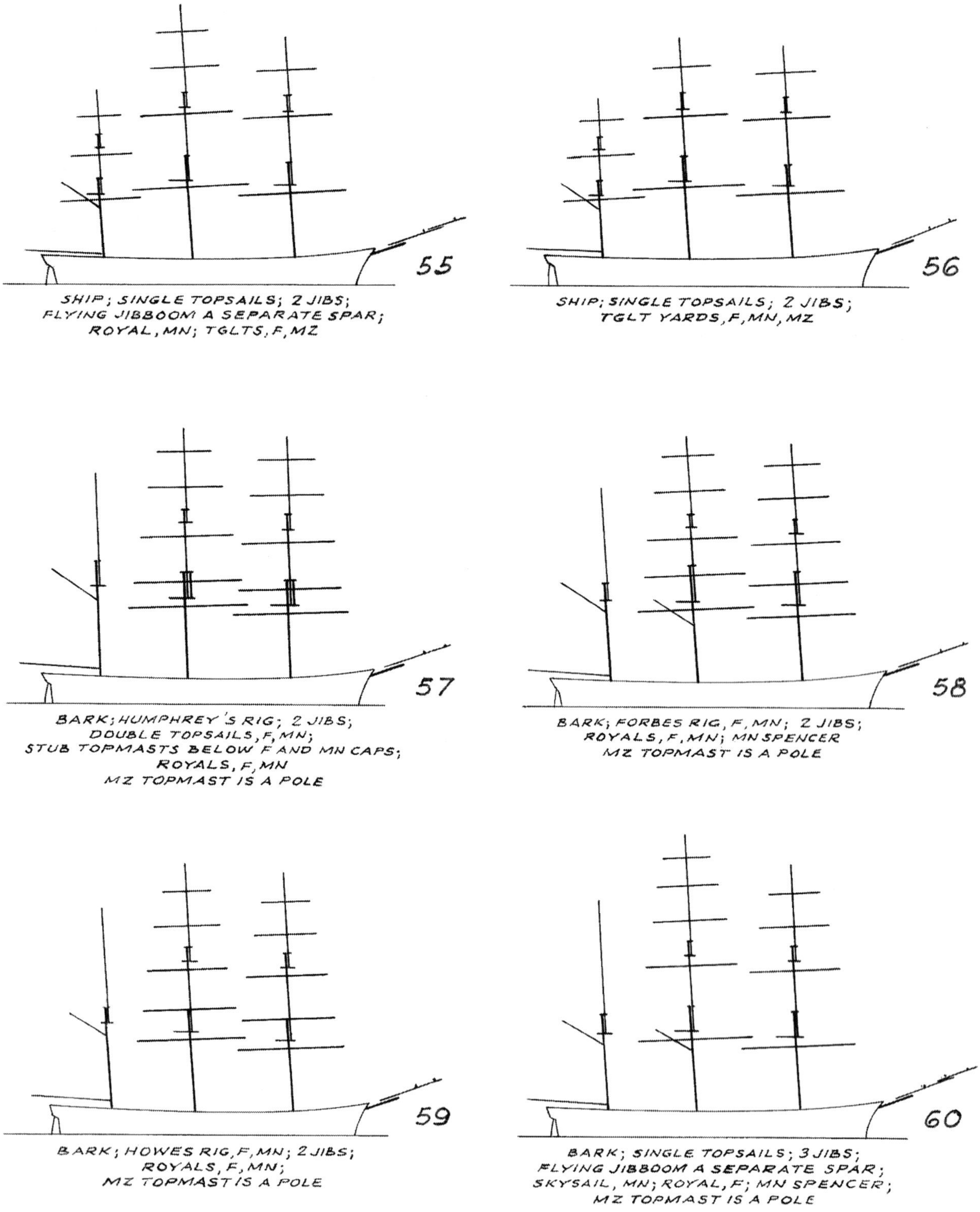

Figure 11.11. Spar plans of American merchant sail, 1850s.

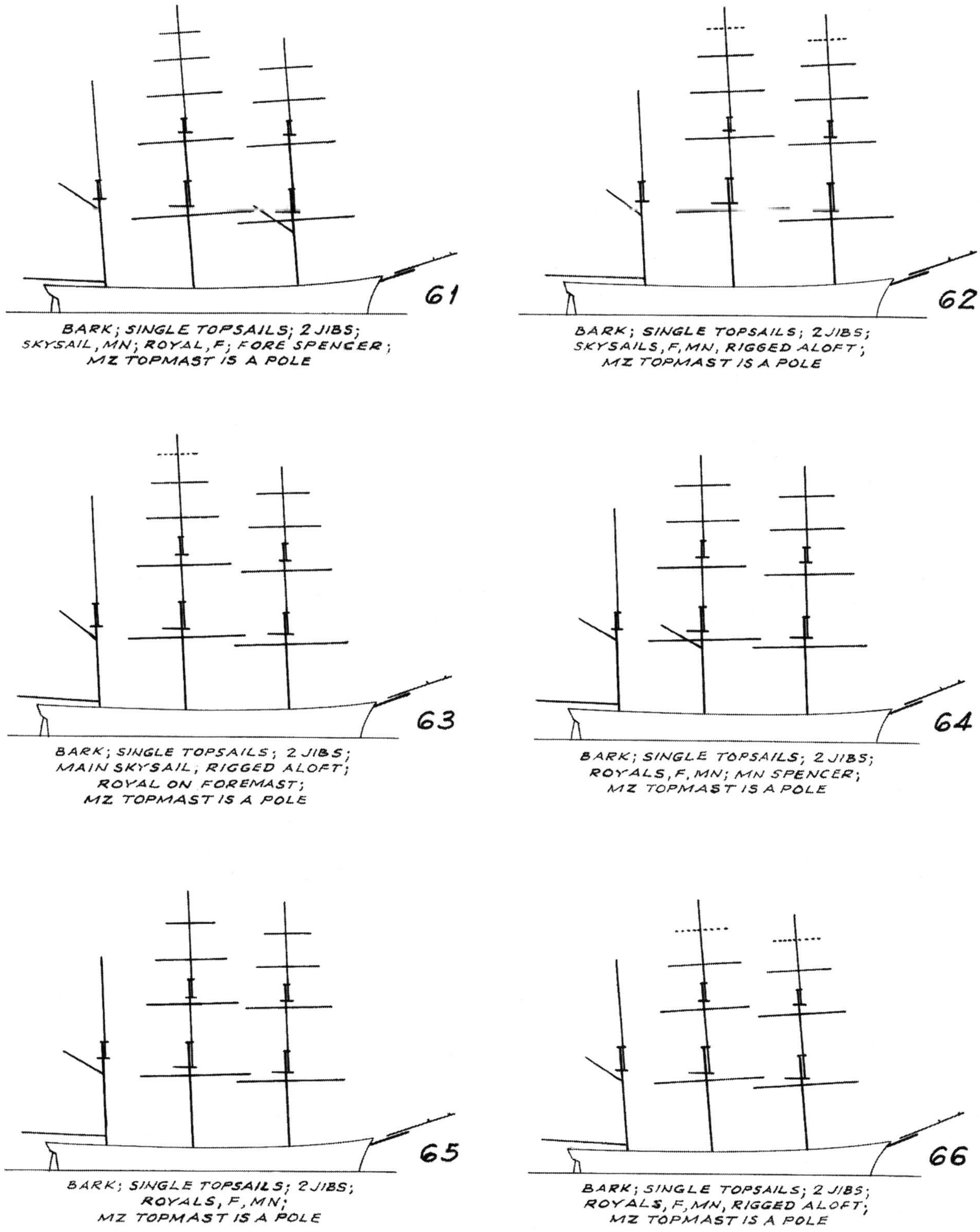

Figure 11.12. Spar plans of American merchant sail, 1850s.

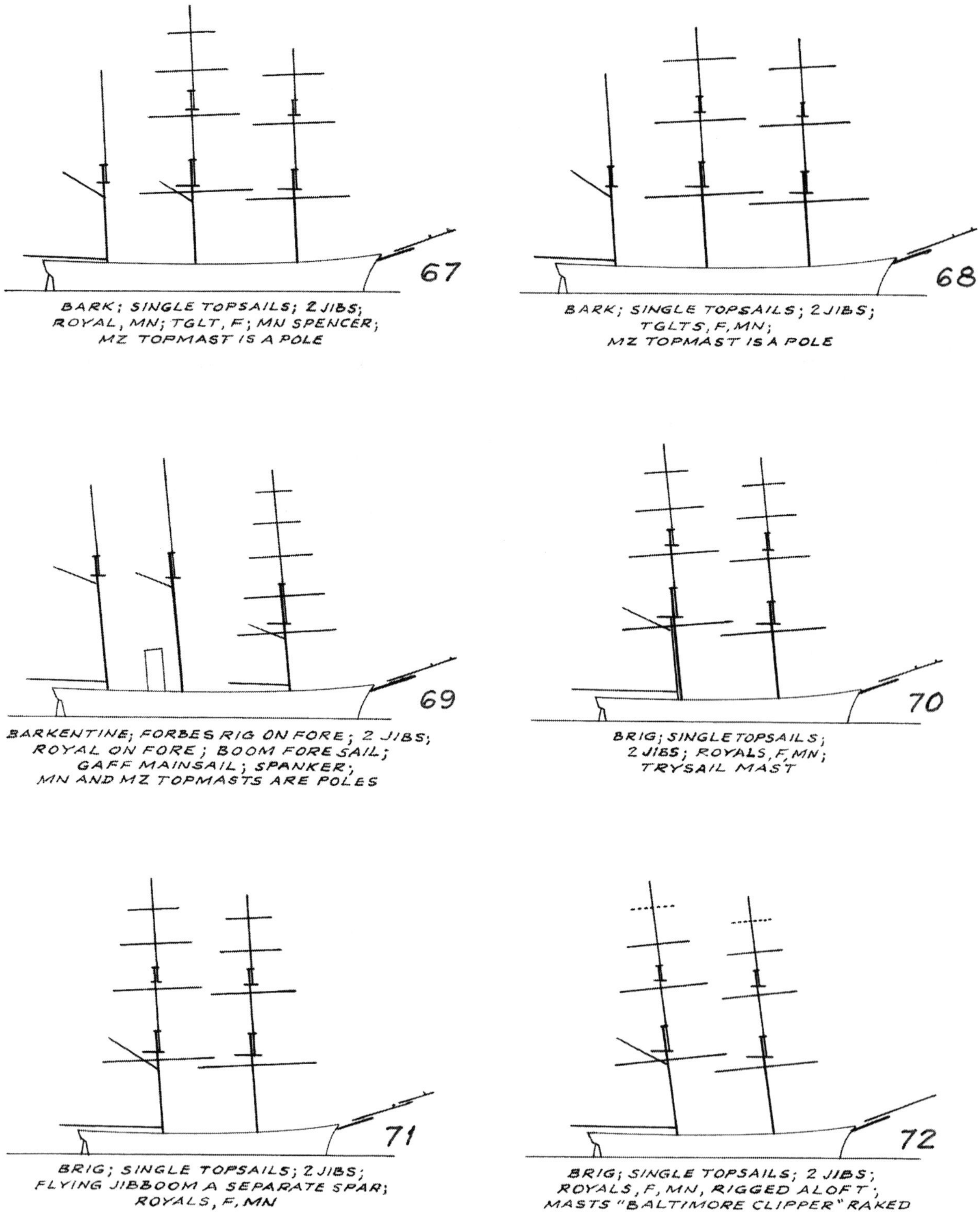

Figure 11.13. Spar plans of American merchant sail, 1850s.

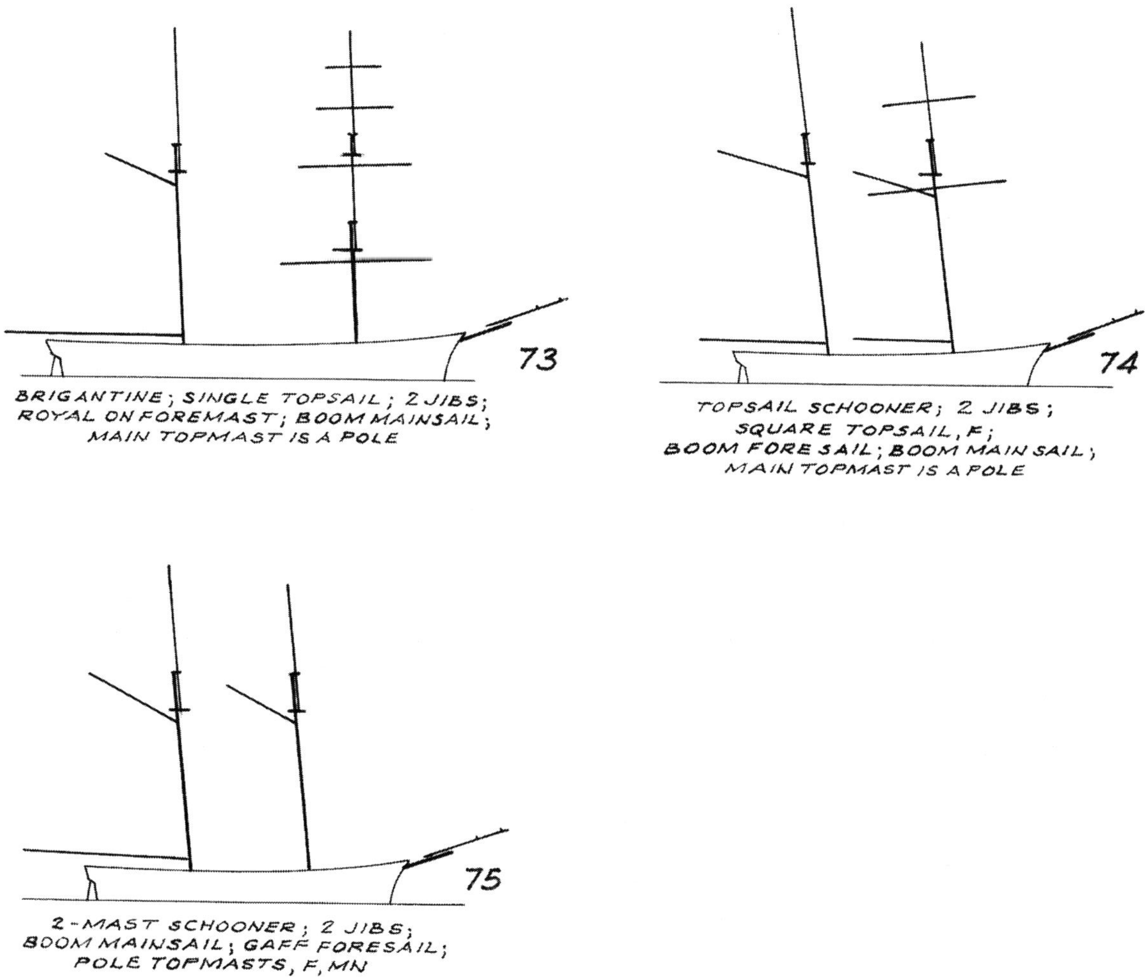

Figure 11.14. Spar plans of American merchant sail, 1850s.

12

The Spar Makers

Craftsmen Unsung in the History of Sail

Throughout the history of the sailing ship endless attention has been devoted to the design and construction of ships' hulls. Development was ever apparent. Hulls were built and refined as experience added its knowledge of the character of the sea to the overall panorama of sea lore.

As ships became larger they required more power in the form of masts and sails to propel them over the vast expanses of the world's oceans.

Over long periods of time the production of both sails and masts became highly specialized endeavors and each was developed to be in accord with the hulls in which they were installed. And as could be expected, each endeavor discovered methods which were peculiar to its own needs.

In the case of masts, many ingenious ways were developed to assemble small trees into large masts when trees of sufficient size were not available. These made assemblages were required to possess adequate strength to accomplish the task at hand.

Complicated seams and joints were invented to build one large unit out of small parts. Once assembled and bound together with iron bands these made masts were generally stronger than single tree masts.

As progress was being made in this manner there was no great effort to record the details of the construction except among those who were engaged in the actual fabrication of such spars.

Generally speaking, all masts appear to be similar in form. They are plain sticks having no outstanding characteristics and do not inspire the urge to write about them. However, an Englishman named John Fincham wrote a book, the *Outline of Shipbuilding*, and included in it a short course in mast making for the use of students of the Royal Naval College. The year was 1829.

At that time he wrote, "As we have no treatise at present in our language at all applicable to the present state of the art of Mast-making, and it is now more than 35 years since any work has been published on this subject [Steel's *Elements of Mastmaking, Sailmaking and Rigging*, 1794], it may be unnecessary to offer any remarks on the attempt to supply the deficiency; and it is to be presumed that, if the changes which have taken place in the practical operations of Mast-making are not sufficient to establish the utility of this undertaking, the omission from every work in English, of the principles on which the masting of ships depends, is a sufficient vindication of the present effort.... The masting of merchant vessels has been entirely omitted; but it is intended shortly to give, in an Appendix, an outline of masting of the merchant navy." This advertisement to the first edition concluded with the sub-note "School of Naval Architecture, July 1829."[39]

The first edition proved to be an outstanding success and was followed by a second edition which included masting of merchant vessels and presented several new methods of combining made-masts. The advertisement to the second edition concluded with the sub-note "Chatham Dockyard, May 1843."

The second edition proved to be a greater success than its predecessor and resulted in an updated third (and final) edition which concluded with the sub-note "Landport, January 1854."

The advertisement to the third edition concludes with the statement that "the Work now contains instruction for the whole class of duties connected with fitting the Masts and Yards to all descriptions of Ships in the Royal Navy, as well as to most descriptions of Vessels in the private service."

The third edition is the one that is used as the primary source of information contained in this book.

I have not embellished the contents in any manner. The illustrations are reconstructed so as best to present the appropriate details.

Fincham's book apparently filled all the needs that a mast maker might require. In the ensuing years, as the wooden sailing ship became ever more obsolete, the need for huge made masts diminished and improvements in their fabrication were no longer relevant. Today the value of the book is measured in its usefulness as a historical work.

The present book attempts to combine the generic contents of Fincham's book with actual American ships that were built and sailed in that era.

The dearth of sources of detailed information relative

to making masts is matched only by the absence of recognition of the men who were the producers of ships' masts. However, posterity was rewarded. In the period between 1850 and 1857 a marine reporter named Duncan McLean, who wrote daily descriptions in profuse detail, reported on ships that were being built in the Boston area. The paper was *The Boston Daily Atlas*.[1] From these descriptions it was possible, literally, to reconstruct the subject vessel. Each description contained the names of her designer and builder and, in some cases, the names of various trades-people.

During this period McLean described approximately 200 vessels. Unfortunately, however, only about a dozen spar makers are specifically identified.

It goes without saying that spar makers were located in, or close to, seaboard cities in order to satisfy the needs of local shipbuilders. However, no source of such information is easily found.

Table 12.1 has been compiled from the available data. It contains the names of spar makers, their locations, the known ships that were built and the dates of building.

In some cases the names are listed as individuals, while in others several individuals operated as companies. Thus, some names, such as Blanchard, Fernald and Pigeon, appear as multiple entries.

Unfortunately, spar making is a subject which has generated little historical interest even though the details included in this work present many ingenious and interesting methods of fabricating a mast.

To conclude this book the masting tables representing all (75) spar plans shown in Chapter 11 have been included in the Appendix. Some of the tables are complete in their content while others consist only of bits of information that are of curious or unusual content.

It is hoped that this book will fill, at least in part, the slot left vacant for masting in the study of wooden sailing vessels.

Table 12.1.
Some New England Spar Makers of the 1850s

Spar Maker	*Location*	*Vessels Masted*	*Date*
Blanchard, Mr.*	Charlestown, Mass.	*Ellen Foster*	1852
Blanchard & Caldwell	Chelsea, Mass.	*Witchcraft*	1851
	"	*Shooting Star*	1851
	"	*John Wade*	1851
	"	*Antelope*	1851
	"	*Winged Arrow*	1852
	"	†*Harry Bluff*	1855
	Charlestown, Mass.	*Golden Light*	1853
	"	*Ocean Pearl*	1853
	"	†*Hesperus*	1856
Fernald, Wm. D.	Boston	*John Bertram*	1851
Fernald & Brother	East Boston	*Hoogly*	1852
Fernald, A.W. & B.L.	"	*John Gilpin*	1852
	"	*Flying Childers*	1852
	"	*Mystery*	1853
Forbes, R.B. (spar designer)	"	*Amphitrite*	1853
Martin, Wm. & Wm. D. Fernald	Portsmouth, N.H.	*Witch of the Wave*	1851
	"	*J. Montgomery*	1852
McKay, Donald	East Boston	*Westward Ho*	1852
	"	*Empress of the Sea*	1853
Odiorne, Thomas H.	Portsmouth, N.H.	*Webster*	1853
	"	†*Witch of the Wave*	1856
Pigeon, Henry	East Boston	*Game Cock*	1851
	"	*Mermaid*, bark	1851
	"	*Sea Bird*, bark	1851
	Medford, Mass.	*Dauntless*	1852
	East Boston	*Queen of Clippers*	1853
	"	*Edwin Forrest*	1853
Pigeon & Odiorne	Boston	*Thatcher Magoun*	1856
Pigeon & Pool†	East Boston	†*Clara Wheeler*	1850
	"	*Racehorse*	1850
	"	*Surprise*	1850

Spar Maker	*Location*	*Vessels Masted*	*Date*
Porter, E.F. Mayhew & Co.	"	*Mastiff*	1856
Spear, Mr.	"	*Wizard*	1853
Titcomb & Lunt***	Newburyport, Mass.	*Racer*	1851
Young, George E.	East Boston	*Stag Hound*	1850
	"	*Star of Empire*	1853
	"	*Donald McKay*	1855
	"	†*Elvira*	1855
	"	†*Abbott Lawrence*	1855
	"	†*Dione*	1857

Data is from *The Boston Daily Atlas*, 1850–1857, except as noted.

†Not a clipper.

*"Mr. Blanchard was known as one of the best spar makers." *The Boston Daily Atlas*, 19 May 1852.

**"They have made the spars for most of Capt. Forbes' ships and also those owned by W.W. Goddard, Esq." *The Boston Daily Atlas*, 5 January 1850.

***"Titcomb & Lunt have been engaged in the business [spar making] in Newburyport for the last fifteen years which is a sufficient recommendation that they are masters of their business." *The Boston Daily Evening Traveler*, 23 June 1851.

Appendix: Masting Tables for Categories of Vessels Included in This Book

This appendix contains a representative masting table for each category of spar plan that is illustrated in Chapter 11 (Figures 11.2 to 11.14 inclusive; category one through category 75). In all cases the number assigned to the masting table (e.g., Spar Plan #1) is the same as the number assigned to the illustration.

The amount of information contained in each table varies widely, ranging between fully complete to some instances where the information consists of a few significant or explanatory sentences.

Masts and Spars

NOTES

1. "—" in the topgallant and royal masthead column indicates a stop, or shoulder, for the rigging.
2. Entries are taken verbatim from the appropriate sources.
3. Items left blank indicate the absence of specific data.
4. All vessels are ship rigged unless otherwise noted.
5. Made masts are those which are composed of a group of sticks. Single tree masts are of one piece of timber.
6. Lower masts are whole lengths unless otherwise noted.
7. In many cases multiple sources are not in agreement.

Appendix, Table 1. Dimensions as Shown in These Lists

Name of Vessel, Year, Ref. # Rig ("ship" unless noted)				***Single or Double Topsails, or as Noted***			
Mast	*Dia., inches*	*Length, feet*	*Head or pole, feet*	*Yard*	*Dia., inches*	*Length, feet*	*Arm, feet, not included in given length*[15,52]
Fore,	at partners	each, heel to head, inclusive		Fore	at the slings, all yards		Does not include yardarms, all yards
Topmast,	at cap			Topsail			
Topgallant,	at cap			Topgallant			
Royal,	at stop	one spar, plus pole		Royal			
Skysail,	at stop			Skysail			
Moonsail,	at stop			Moonsail			
Main,	at partners	each, heel to head, inclusive		Main			Does not include yardarms, all yards
Topmast,	at cap			Topsail			
Topgallant,	at cap			Topgallant			
Royal,	at stop	one spar, plus pole		Royal			
Skysail,	at stop			Skysail			
Moonsail,	at stop			Moonsail			
Mizzen,	at partners	each, heel to head, inclusive		Crojack			Does not include yardarms, all yards
Topmast,	at cap			Topsail			
Topgallant,	at cap			Topgallant			
Royal,	at stop	one spar, plus pole		Royal			
Skysail,	at stop			Skysail			
Moonsail,	at stop			Moonsail			

Name of Vessel, Year, Ref. # Rig ("ship" unless noted)				*Single or Double Topsails, or as Noted*			
Mast	*Dia., inches*	*Length, feet*	*Head or pole, feet*	*Yard*	*Dia., inches*	*Length, feet*	*Arm, feet, not included in given length*[15,52]
Bowsprit,	at stem	stem, outboard to cap		Spanker boom	max.*		
Jibboom,	at cap	outside of cap		Spanker gaff	at heel	includes end	
Flying jibboom		outside of jib, plus end		Spencer gaffs			

Notes

*Diameter regardless of location on spar.
Exceptions are noted in the appropriate lists.
The stop is a shoulder.

Spar Plan #1

Great Republic, 1853 #2, 6, 7 — clipper
4-masted bark — double topsails,
(original) — Forbes rig

Mast	*Dia., inches*	*Length, feet*	*Head or pole, feet*	*Yard*	*Dia., inches*	*Length, feet*	*Arm, feet*
Fore	44	130	36	Fore	26	110	6
Topmast	24	76	12	Lower topsail	24	90	5
				Upper topsail	19	76	4½

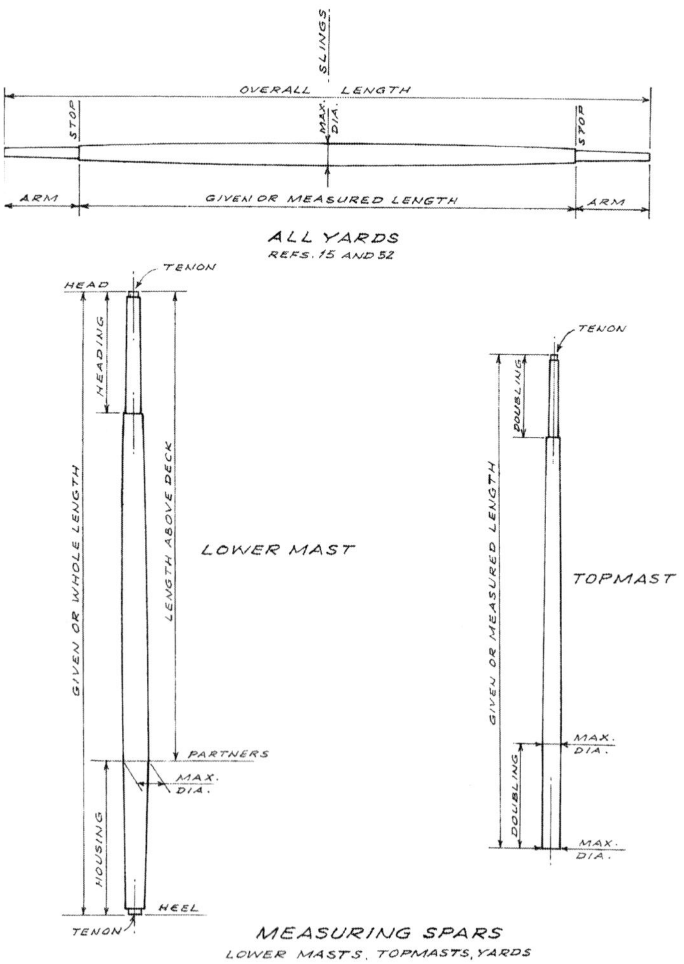

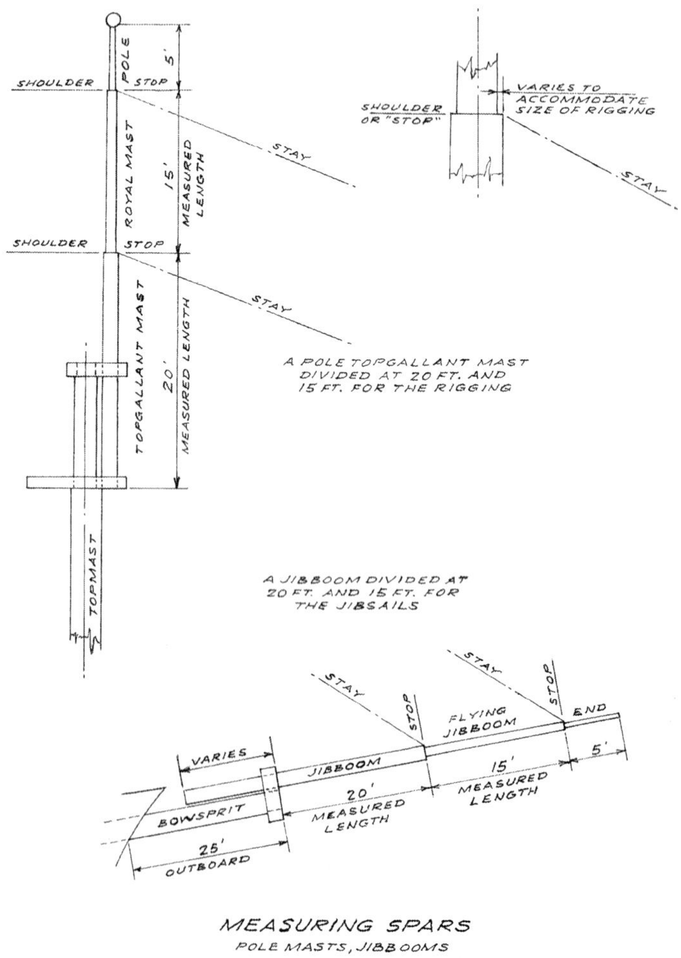

Left: Measuring spars: lower masts; topmasts; yards. *Right:* Measuring spars: pole masts; jibbooms.

Mast	*Dia., inches*	*Length, feet*	*Head or pole, feet*	*Yard*	*Dia., inches*	*Length, feet*	*Arm, feet*
Topgallant	18	28	—	Topgallant	15	62	4
Royal	15	22	—	Royal	12	51	3½
Skysail	11	19	12, pole	Skysail	9	40	3
Main	44	131	36	Main	28	120	6
Topmast	24	76	12	Lower topsail	24	92	5
				Upper topsail	19	76	4½
Topgallant	18	28	—	Topgallant	15	62	4
Royal	15	22	—	Royal	12	51	3½
Skysail	11	19	12, pole	Skysail	9	40	3
Mizzen	40	122	33	Crojack	24	90	5
Topmast	22	69	10	Lower topsail	19	76	4½
				Upper topsail	15	62	4
Topgallant	16	22	—	Topgallant	12	51	3½
Royal	10	19	—	Royal	9	40	3
Skysail	8	12	8, pole	Skysail	6	29	2
Spanker mast	26	110	14				
Topmast		40	7, pole*				
Bowsprit	44	30, outboard		Spanker boom		40	2, end
Inner jibboom	22	18, outside	4, end	Spanker gaff		34	8, end
Outer jibboom		14					
Flying jibboom		8	6, end				

*A pole mast divided at 15 feet and 10 feet above the cap for the gaff topsail and gaff topgallant sail.

NOTES

1. Masts are square rigged, except spanker mast, which is fore-and-aft rigged.
2. Fore, main and mizzen masts and bowsprit are made, doweled together, bolted and hooped over all, including mastheads, with iron.
3. The spanker topmast is a pole mast.
4. All masts and jibbooms are of hard pine.
5. Spacing of masts, measured on deck, ref. 7. Stem to foremast, 76 ft.; foremast to mainmast, 103 ft.; mainmast to mizzenmast, 85 ft.; mizzenmast to spanker mast, 46 ft.; spanker mast to taffrail, 25 ft.
6. Masting data in refs. 6 and 7 are not in complete agreement. These dimensions are take from ref. 7.
7. Masts rake, ½" fore, main and mizzen, per foot; 1½" spanker, per foot.

Spar Plan #2

Great Republic, (rebuilt, 1855) #2, 8 4-masted bark	clipper double topsails, Forbes rig

Mast	*Dia., inches*	*Length, feet*	*Head or pole, feet*	*Yard*	*Dia., inches*	*Length, feet*	*Arm, feet*
Fore	41	90	18	Fore	24	85	5
Topmast	20½	53	10	Lower topsail	20	76	4½
				Upper topsail	18	67	5
Topgallant		29	—	Topgallant	13½	49	3½
Royal		18½	—	Royal	10½	38	2½
Skysail		13	10, pole	Skysail	7	27	1½
Main	41	95	18	Main	24	90	5
Topmast	20½	55	10	Lower topsail	20	81	4½
				Upper topsail	18	72	5
Topgallant		29	—	Topgallant	14½	54	3½
Royal		18½	—	Royal	11½	43	2½
Skysail		13	10, pole	Skysail	8	32	1½
Mizzen	34	86	14	Crojack	20	72	4½
Topmast	17	45	9	Lower topsail	16	63	4½

Mast	Dia., inches	Length, feet	Head or pole, feet	Yard	Dia., inches	Length, feet	Arm, feet
				Upper topsail	14½	55	4
Topgallant		23	—	Topgallant	11½	43	2½
Royal		13½	—	Royal	8	32	2
Skysail		10	9, pole	Skysail	5¼	21	1
Spanker mast	28	68	12				
Topmast		72	7, pole*				
Bowsprit	44	30, outboard		Spanker boom		42	2, end
Inner jibboom	22	18, outside		Spanker gaff		36	6, end
Outer jibboom		14					
Flying jibboom		8	6, end				

*A pole mast divided at 26 feet, 20 feet, and 14 feet above the cap for the gaff topsail and stops for the rigging.

Notes

Mast spacing is same as original ship, q.v.
Masts are square rigged, except spanker mast, which is fore-and-aft rigged.

Spar Plan # 3

Seaman's Bride, 1851 #6 — clipper, single topsails

Mast	Dia., inches	Length, feet	Head or pole, feet	Yard	Dia., inches	Length, feet	Arm, feet
Fore				Fore			
Topmast				Topmast			
Topgallant				Topgallant			
Royal				Royal			
Skysail				Skysail			
Moonsail				Moonsail			
Main				Main		68	
Topmast		40		Topmast			
Topgallant				Topgallant			
Royal				Royal			
Skysail				Skysail			
Moonsail				Moonsail			
Mizzen				Crojack			
Topmast				Topmast			
Topgallant				Topgallant			
Royal				Royal			
Skysail				Skysail			
Moonsail				Moonsail			
Bowsprit				Spanker boom			
Jibboom				Spanker gaff			
Flying jibboom							

Notes

All lower masts are single spars.
Ref. 6 (*New York Herald*) on 29 June 1851 reports "her sail plan showed that she was one of very few ships to carry three moonsails above her skysails."
A suit of sails will spread nearly 7,000 yards of canvas.
No additional details recorded.

Spar Plan #4

Hurricane, 1851 #2, 6 — clipper, single topsails

Mast	Dia., inches	Length, feet	Head or pole, feet	Yard	Dia., inches	Length, feet	Arm, feet
Fore	34	85	14½	Fore	20½	80	5
Topmast	17	48½	8½	Topmast	15½	62	5
Topgallant	12	25	—	Topgallant	11	44	3½
Royal		15½	—	Royal	8	33½	2¼
Skysail		10'-5"	5' pole	Skysail	6¾	27	1¾
Moonsail		*		Moonsail		25*	
Main	35	89	14½	Main	20½	80	5
Topmast	17½	50½	8½	Topsail	15½	62	5
Topgallant	12	26½	—	Topgallant	11	44	3½
Royal		17	—	Royal	8	33½	2¼
Skysail		12½	—	Skysail	6¾	27	1¾
Moonsail		9	5' pole	Moonsail		25*	
Mizzen	25	80	11½	Crojack	15½	60	4½
Topmast		41	7¼	Topsail	10¾	45	3½
Topgallant		19½	—	Topgallant	7¼	30	2½
Royal		13½	—	Royal	5	21¾	1¾
Skysail		9	3½, pole	Skysail	4	16	1¼
Bowsprit				Spanker boom			
Jibboom				Spanker gaff			
Flying jibboom							

NOTES

Lower masts and bowsprit are made.

*A discrepancy exists in moonsail data. No moonmast length is recorded for the foremast in ref. 6.

All yard lengths for fore and mainmasts are listed under a combined heading.

Other spars are in proportion.

Mast spacing: Knightheads to center of foremast, 52½ ft.; thence to center of mainmast, 72½ ft.; thence to center of mizzenmast, 57½ ft.; thence to taffrail, 45 ft.

Ref. 2 states that *Hurricane* was fitted with Cunningham's rolling topsails. These sails did not appear until 1856.

Spar Plan #5

Young America, 1853 #2, 4, 16 — clipper, single topsails

Mast	Dia., inches	Length, feet	Head or pole, feet	Yard	Dia., inches	Length, feet	Arm, feet
Fore	36	92	16½	Fore	22	82	4½
Topmast	18	53	9	Topmast	16¾	69⅔	6-2 in.
Topgallant*	13	36	6	Topgallant	11¾	51	2¾
Royal	8½	20	—	Royal	8¾	40	1-7 in.
Skysail	5½	13	10⅔, pole 3¾ in. dia.	Skysail	6¼	30½	0-9 in.
Main	36	96	17	Main	23	86	5
Topmast	19	56	9½	Topsail	17¾	73	7-2 in.
Topgallant*	13½	37¾	6⅓	Topgallant	12 ½	52⅔	3⅓
Royal	9	20¾	—	Royal	9	41⅓	1⅔
Skysail	6	14¼	—	Skysail	6½	31	1
Moonsail	4	9½	7¾, pole 4 in. dia.	Moonsail	4½	24½	0-6 in.

Mast	Dia., inches	Length, feet	Head or pole, feet	Yard	Dia., inches	Length, feet	Arm, feet
Mizzen	29	84½	13½	Crojack	17¾	68	3½
Topmast	15	44	7⅓	Topsail	14	57	5
Topgallant*	10¼	29	5⅓	Topgallant	9½	41⅔	2
Royal	6¾	16	—	Royal	6¾	32⅓	1½
Skysail	4½	10¾	7½, pole 3 in. dia.	Skysail	5	25	0-6 in.
Bowsprit	33 abt.	28, outboard		Spanker boom	14	62	3, end
Inner jibboom	17½	18, outside		Spanker gaff	9	40	6½, end
Outer jibboom		14	5, end	Main spencer gaff		26	2, end
				Fore spencer gaff		26	2, end
				Swinging booms		64	
				Monkey Gaff	5	22	

NOTES

These dimensions are taken from a contemporary masting list which is in general agreement with the sail plan of *Young America* ref. 4.
*Topgallant mast is a separate spar, all masts.
Fore top is 17-9 in. wide; main top 18-6 in. wide; mizzen top 14-6 in. wide. Royal masts and above are a single spar, all masts.
Main yard is scarphed in the bunts and banded with 12 iron bands, ref. 16.

Spar Plan #6 (or 7)

Belle of the West, 1853 #1, 2 — clipper, single topsails, Forbes rig. See notes below.

Mast	Dia., inches	Length, feet	Head or pole, feet	Yard	Dia., inches	Length, feet	Arm, feet
Fore				Fore			
Topmast				Lower topsail			
				Upper topsail			
Topgallant				Topgallant			
Royal				Royal			
Skysail				Skysail			
Main		85	23	Main		72	
Topmast		50		Lower topsail		63	
				Upper topsail		63	
Topgallant		20		Topgallant		53	
Royal		14		Royal		40	
Skysail		9		Skysail		31	
Mizzen				Crojack			
Topmast				Lower topsail			
				Upper topsail			
Topgallant				Topgallant			
Royal				Royal			
Skysail				Skysail			
Bowsprit				Spanker boom			
Jibboom				Spanker gaff			
Flying jibboom							

NOTES

Ref. 1, "Topmasts are fitted before the lower mastheads."
Ref. 2, "double topsails, with topmasts fitted abaft the lower mastheads."
(Ref. 1 is an on-site description.)
Mast and yard dimensions are taken from ref. 2. Skysails on all masts are implied in the references.
Other spars are in proportion.
No additional details are recorded.

Spar Plan #7

Amphitrite, 1853 #1 — clipper, double topsails, Forbes rig

Mast	*Dia., inches*	*Length, feet*	*Head or pole, feet*	*Yard*	*Dia., inches*	*Length, feet*	*Arm, feet*
Fore		94	25	Fore		81¼	
Topmast		55½		Lower topsail		69	
				Upper topsail		54½	
Topgallant				Topgallant		43	
Royal				Royal		32	
Skysail				Skysail			
Main		100	26	Main		92	
Topmast		59		Lower topsail		81¼	
				Upper topsail		69	
Topgallant				Topgallant		54½	
Royal				Royal		43	
Skysail				Skysail		32	
Mizzen		97	22	Crojack		69	
Topmast		47		Lower topsail		54½	
				Upper topsail		43	
Topgallant				Topgallant		32	
Royal				Royal			
Skysail				Skysail			
Bowsprit				Spanker boom			
Jibboom				Spanker gaff			
Flying jibboom							

NOTES

"The fore yard and those above are nearly the same as those on the mainmast, commencing at the lower topsail yard, and the same is true of the mizzenmast, commencing a yard higher up; thus, excepting the main yard, she has duplicate yards on the other masts. Her spars were draughted by Capt. Forbes himself."

Topmasts are fidded before the lower mastheads.

Fore and main lower masts are made, and hooped over all. Mizzenmast is a single spar.

Topmasts and jibbooms are of hard pine.

Other masts and spars are in proportion.

No additional details are recorded.

Spar Plan #8

Thatcher Magoun, 1856 #1, 2 — clipper, double topsails, Howes rig

Mast	*Dia., inches*	*Length, feet*	*Head or pole, feet*	*Yard*	*Dia., inches*	*Length, feet*	*Arm, feet*
Fore	32	83	14½	Fore	21	80	5
Topmast	18	46	8½	Lower topsail	15	68½	4
				Upper topsail	13	59	4
Topgallant	12	23	—	Topgallant	12	48	3½
Royal	10	16½	—	Royal	9	38½	2½
Skysail	8½	12	8, pole	Skysail	7	32	2
Main	33	86	14½	Main	21	80	5
Topmast	18	46	8½	Lower topsail	15	68½	4
				Upper topsail	13	59	4
Topgallant	12	23	—	Topgallant	12	48	3½

Mast	*Dia., inches*	*Length, feet*	*Head or pole, feet*	*Yard*	*Dia., inches*	*Length, feet*	*Arm, feet*
Royal	10	16½	—	Royal	9	38½	2½
Skysail	8½	12	8, pole	Skysail	7	32	2
Mizzen	27	77½	11½	Crojack	17	60	4
Topmast	14	38½	6	Lower topsail	13	51½	3½
				Upper topsail	11	43½	3
Topgallant	10	17	—	Topgallant	8½	35½	2½
Royal	8	13½	—	Royal	6½	27½	2
Skysail	6	9½	8, pole	Skysail	4½	20	1½
Bowsprit	30½	22, outboard		Spanker boom			
Jibboom	17½	18, outside		Spanker gaff			
Flying jibboom		15	5, end				

NOTES

Fore and main lower masts and bowsprit are made and are of hard pine.
Topmasts and jibbooms are of hard pine.
Other spars are in proportion.

Spar Plan #9

Flying Childers, 1852 #1 — clipper, double topsails, Forbes rig

Mast	*Dia., inches*	*Length, feet*	*Head or pole, feet*	*Yard*	*Dia., inches*	*Length, feet*	*Arm, feet*
Fore	31	81¼	21	Fore	16½	66	4
Topmast	15	45	6¼	Lower topsail	13¾	57	4
				Upper topsail	10	47	4
Topgallant	9	15¾	—	Topgallant	8½	36	2½
Royal	7	13½	—	Royal	6¼	26	1
Skysail	5	9¼	1, pole	Skysail	4¼	20	0-9 in.
Main	32	87	23	Main	19	78	4
Topmast	15½	50	7½	Lower topsail	14	68	4½
				Upper topsail	12	56	3½
Topgallant	10	16¾	—	Topgallant	9½	46¾	3
Royal	9	15¼	—	Royal	8½	36	2½
Skysail	6½	13	1, pole	Skysail	6¼	26	1
Mizzen	23	84¼	17¾	Crojack	12½	57	4
Topmast	10½	34	—	Lower topsail	10¼	47	4
				Upper topsail	8¼	36	2½
Topgallant	7	13¼	—	Topgallant	6	26	1
Royal	5	9	1, pole	Royal	4¾	20	0-9 in.
Bowsprit	29	23½, outboard		Spanker boom	10½	50	2, end
Jibboom*	15½	15, outside	3, end	Spanker gaff	7	26	3, end
Flying jibboom				Trysail mast			

NOTES

*Jibboom is 11 feet inboard of bowsprit cap.

Mizzen topmast and above are a pole mast.

The fore and mainmasts are fitted abaft the lower mastheads. Mizzen topmast is fitted before the mizzen masthead to avoid interference with the trysail mast, in sending the topmast up or down.

"She has iron bands over her trestletrees, bolted through them; and, in the wake of the mastheads, an iron band on each side, fitted over each trestle tree, and also bolted. The other bands extend across, before and abaft the mastheads, and have each an eye to which a pair of shrouds are shackled. The heads of the lower masts project above the caps, and over these projections are iron bands, to which her masthead shrouds and stays are also shackled."

Spar Plan #10

Witch of the Wave, 1856 #1	freighter, double topsails Forbes rig

Mast	*Dia., inches*	*Length, feet*	*Head or pole, feet*	*Yard*	*Dia., inches*	*Length, feet*	*Arm, feet*
Fore	30	79½	14	Fore	19	72	3½
Topmast	16	45	9	Lower topsail	15	61	3
				Upper topsail	15	56	3
Topgallant	11½	25	0	Topgallant	11	42	2½
Royal	10	17	9, pole	Royal	7	32	2
Main	31	83	14	Main	19	76	3½
Topmast	18	45½	9, pole	Lower topsail	16	65	3
				Upper topsail	15	57	3
Topgallant	11½	25½	0	Topgallant	11	42½	2½
Royal	10	17½	0	Royal	7½	32	2
Skysail, main only		*	11, pole	Skysail, main only		*	
Mizzen	26	75	11½	Crojack	15	56	3
Topmast	18	36	7	Lower topsail	12	50	2½
				Upper topsail	11	43	2½
Topgallant	9	19	0	Topgallant	8½	32	2
Royal	7	13½	7, pole	Royal	6	24	1
Bowsprit	30	26, outboard		Spanker boom		49	
Jibboom		18, outside		Spanker gaff		38	
Flying jibboom		14	6, end	Main spencer gaff		16	

NOTES

Masts rake, ¾" fore; ⅞" main; 1¼" mizzen, per foot.
Foremast, mainmast and bowsprit are made.
Topmasts and jibbooms are hard pine.
*No dimensions given.

Spar Plan #11

Donald McKay, 1855 #1	clipper, double topsails Howes rig

Mast	*Dia., inches*	*Length, feet*	*Head or pole, feet*	*Yard*	*Dia., inches*	*Length, feet*	*Arm, feet*
Fore	41	72	18	Fore	26	105	5
Topmast	21	54	11	Lower topsail	24	87	3
				Upper topsail	21	80	5½
Topgallant	15	28	—	Topgallant	15	59	3½
Royal	13	18	9, pole	Royal	11	46	2½
Main	42	77	18	Main	27	115	5
Topmast	22	56	11	Lower topsail	25	96	3
				Upper topsail	22	86	5½
Topgallant	15	28	—	Topgallant	15	64	3½
Royal	13	18	—	Royal	11	51	3
Skysail	9	12	9, pole	Skysail	9	40	1½
Mizzen	32	67	15	Crojack	22	82	5
Topmast	18	46½	9, pole	Lower topsail	19	69	3
				Upper topsail	17	64	4½
Topgallant	13	26	—	Topgallant	15	48	3
Royal	11	16	8, pole	Royal	10	38	2

Mast	Dia., inches	Length, feet	Head or pole, feet	Yard	Dia., inches	Length, feet	Arm, feet
Bowsprit	40	22, outboard		Spanker boom		58	
Inner jibboom	22	15, outside		Spanker gaff		44	
Outer jibboom		14					
Flying jibboom		10	5, end				

NOTES

Lower masts and bowsprit are made, are of pitch pine, and are hooped with iron. Hoops on fore and main masts are four inches wide by ⅝ inch thick; 30, and 31 in number. Mizzenmast has 28 hoops, no size given.

Fore and main lower yards are hooped with iron in the bunts.

Jibboom, topmasts, and lower yards are of pitch pine.

Lower studdingsails are triangular, requiring no swinging booms.

Square sails and fore-and-aft sails spread 14,159 yards of canvas.

Masts are nearly upright.

Spar Plan #12

Golden Fleece, 1855 #1 — clipper, double topsails, Howes rig

Mast	Dia., inches	Length, feet	Head or pole, feet	Yard	Dia., inches	Length, feet	Arm, feet
Fore		84		Fore		75	
Topmast		47		Lower topsail		63½	
				Upper topsail		55½	
Topgallant		26		Topgallant		45	
Royal		17	8, pole	Royal		36	
Main		88		Main		81	
Topmast		48		Lower topsail		69	
				Upper topsail		64	
Topgallant		27		Topgallant		50	
Royal		18		Royal		40	
Skysail		12	6, pole	Skysail		32	
Mizzen		78		Crojack		61	
Topmast		40		Lower topsail		50	
				Upper topsail		44½	
Topgallant		22		Topgallant		36	
Royal		14	6, pole	Royal		28	
Bowsprit		20, outboard		Spanker boom			
Jibboom		16, outside		Spanker gaff			
Flying jibboom		13	6, end				

NOTES

Fore and main lower masts and bowsprit are made.

Topgallant masts and above are pole masts.

Topmasts are of hard pine.

Masts and spars are in proportion.

No additional details are recorded.

Spar Plan #13

Climax, 1853 #1, 2	clipper, double topsails Howes rig

Mast	*Dia., inches*	*Length, feet*	*Head or pole, feet*	*Yard*	*Dia., inches*	*Length, feet*	*Arm, feet*
Fore	30	74	12¼	Fore	17	68	3½
Topmast	15	42	7½	Lower topsail	14½	61	3⅙
				Upper topsail	14½	54	3
Topgallant	11	24	—	Topgallant	10½	40	2½
Royal	10	13	—	Royal	7½	29	2
Skysail	8	11	no pole	Skysail	6	19	1½
Main	30	79	13	Main	18½	72	2½
Topmast	15½	44	7½	Lower topsail	15	65	3⅙
				Upper topsail	14½	58	3
Topgallant	11	25	—	Topgallant	10½	42	2½
Royal	10	14	—	Royal	7½	31	2
Skysail	8	12	9, pole	Skysail	6	21	1⅙
Mizzen	24	70	10	Crojack	14	53	3
Topmast	12	32	6	Lower topsail	11	47	2¾
				Upper topsail	11	42	2¼
Topgallant	8	18	—	Topgallant	8	32	2
Royal	7	11	—	Royal	6	22	1¾
Skysail	6	9	no pole	Skysail	5	15	1¼
Bowsprit		25, outboard		Spanker boom			
Jibboom		18, outside		Spanker gaff			
Flying jibboom		14	6, end				

NOTES

Climax was the first vessel to be fitted with Howes rig, ref. 2. Topmasts are fidded before the lower mastheads; lower topsail yards are stationary, being trussed to the fore part of the caps.

All of *Climax*'s skysail yards are set flying, that is, rigged aloft.

Spar Plan #14

Sword Fish, 1851 #2	clipper, double topsails Howes rig, on fore and main only

Mast	*Dia., inches*	*Length, feet*	*Head or pole, feet*	*Yard*	*Dia., inches*	*Length, feet*	*Arm, feet*
Fore				Fore			
Topmast				Lower topsail			
				Upper topsail			
Topgallant				Topgallant			
Royal				Royal			
Skysail				Skysail			
Main				Main			
Topmast				Lower topsail			
				Upper topsail			
Topgallant				Topgallant			
Royal				Royal			
Skysail				Skysail			
Mizzen				Crojack			
Topmast							

Mast	*Dia., inches*	*Length, feet*	*Head or pole, feet*	*Yard*	*Dia., inches*	*Length, feet*	*Arm, feet*
Topsail (See note below)							
Topgallant				Topgallant			
Royal				Royal			
Skysail				Skysail			
Bowsprit				Spanker boom			
Jibboom				Spanker gaff			
Flying jibboom							

NOTES

A painting in ref. 2 shows the vessel with Howes rig on the fore and mainmasts only, apparently a re-rig, date unknown.

Spar Plan #15

Blue Jacket, 1854 #2	clipper, double topsails Howes rig, modified

Mast	*Dia., inches*	*Length, feet*	*Head or pole, feet*	*Yard*	*Dia., inches*	*Length, feet*	*Arm, feet*
Fore				Fore			
Topmast				Lower topsail			
				Upper topsail			
Topgallant				Topgallant			
Royal				Royal			
Main				Main			
Topmast				Lower topsail			
				Upper topsail			
Topgallant				Topgallant			
Royal				Royal			
Skysail, main only. See note.					Skysail, main only, See note.		
Mizzen				Crojack			
Topmast				Topsail			
Topgallant				Topgallant			
Royal				Royal			
Bowsprit				Spanker boom			
Jibboom				Spanker gaff			
Flying jibboom							

NOTES

Blue Jacket crossed double topsail yards on fore and main masts only. Mizzen was fitted with a single topsail.
Painting in ref. 2 shows a skysail on mainmast only. Description states that she did not cross skysail yards.
No details are recorded.

Spar Plan #16

Cornelius Grinnell, 1850 #28	packet, double topsails Forbes rig

Mast	*Dia., inches*	*Length, feet*	*Head or pole, feet*	*Yard*	*Dia., inches*	*Length, feet*	*Arm, feet*
Fore				Fore			
Topmast				Lower topsail			
				Upper topsail			

Mast	Dia., inches	Length, feet	Head or pole, feet	Yard	Dia., inches	Length, feet	Arm, feet
Topgallant				Topgallant			
Royal				Royal			
Skysail				Skysail			
Main				Main			
Topmast				Lower topsail			
				Upper topsail			
Topgallant				Topgallant			
Royal				Royal			
Skysail				Skysail			
Mizzen				Crojack			
Topmast				Lower topsail			
				Upper topsail			
Topgallant				Topgallant			
Royal				Royal			
Skysail				Skysail			
Bowsprit				Spanker boom			
Jibboom				Spanker gaff			
Flying jibboom							

NOTES

No mast dimensions are available. An article in *The Boston Daily Atlas*, 17 January 1855, states that the ship was fitted with Forbes rig with the topmasts fidded abaft the lower masts (Ref. 28).

Spar Plan #17

Flying Mist, 1856 #1, 2 — clipper, double topsails, Linnell's rig

Mast	Dia., inches	Length, feet	Head or pole, feet	Yard	Dia., inches	Length, feet	Arm, feet
Fore	33	76	14	Fore	20	74	3½
Topmast	16½	41	8	Lower topsail	17	65	3
				Upper topsail	15	56	3
Topgallant	12	24	—	Topgallant	12	44	2½
Royal	9	15	7, pole	Royal	8	36	2
Main	34	80	14	Main	20	74	3½
Topmast	16½	41	8	Lower topsail	17	65	3
				Upper topsail	15	56	3
Topgallant	12	24	—	Topgallant	12	44	2½
Royal	9	15	—	Royal	8	36	2
Skysail	7	12	7, pole	Skysail*	6	30	1
Mizzen	26	72	12½	Crojack	15	60	2½
Topmast	13	37	6½	Lower topsail	12	52½	2
				Upper topsail	12	45	2½
Topgallant	9½	20	—	Topgallant	9	34	1½
Royal	7½	13	5, pole	Royal	7	30	1
Bowsprit	34	20, outboard		Spanker boom		47	
Jibboom	16	18, outside		Spanker gaff		42	
Flying jibboom		14	6, end				

NOTES

All lower masts and bowsprit are made.
Topmasts and jibbooms are of hard pine.
*Main skysail yard is rigged aloft.

Spar Plan #18

Hesperus, 1856 #1 — freighter, double topsails, Howes rig

Mast	*Dia., inches*	*Length, feet*	*Head or pole, feet*	*Yard*	*Dia., inches*	*Length, feet*	*Arm, feet*
Fore	31	74	13½	Fore	18	68	3
Topmast	16	44	7½	Lower topsail	14	60	3
				Upper topsail	13	54	3
Topgallant	11½	23½	0	Topgallant	10½	42	2½
Royal	8	15½	0	Royal	8	33	1¼
Main	31	79	13½	Main	18	68	3
Topmast	16	44	7½	Lower topsail	14	60	3
				Upper topsail	13	54	3
Topgallant	11½	23½	0	Topgallant	10½	42	2½
Royal	8	15½	0	Royal	8	33	1¼
Skysail*	6	12½	6, pole	Skysail*	6¼	27	1
Mizzen	25	71	10	Crojack	14½	54	3
Topmast	13½	34	6	Lower topsail	11½	47	2½
				Upper topsail	10	41	2½
Topgallant	8½	18½	0	Topgallant	8	34	1½
Royal	6	12	5, pole	Royal	5½	24½	1
Bowsprit	29	20, outboard		Spanker boom		44	
Jibboom	16	14, outside		Spanker gaff		36	
Flying jibboom		13	5, end	Main spencer gaff		20	

NOTES

Masts rake, ⅞" fore; 1" main; 1¼ mizzen, per foot.
*Main skysail yard is rigged aloft.

Spar Plan #19

Ocean Monarch, 1856 #11 — freighter, double topsails, Howes rig

Mast	*Dia., inches*	*Length, feet*	*Head or pole, feet*	*Yard*	*Dia., inches*	*Length, feet*	*Arm, feet*
Fore	34	87	17	Fore	23	83	4½
Topmast	16	52½	9	Lower topsail	18	74½	3½
				Upper topsail	16	69	5
Topgallant	12	30	—	Topgallant	12	50	3
Royal	11	16	9½	Royal	7	37	2
Main	35	90	17½	Main	23	83	4½
Topmast	17	53	9½	Lower topsail	18	74½	3½
				Upper topsail	16	69	5
Topgallant	12	30½	—	Topgallant	12	50	3
Royal	11	18½	10, pole	Royal	7	37	2
Mizzen	25	81½	14	Crojack	15	67	4
Topmast	13	42	7½	Lower topsail	13	60	3½
				Upper topsail	12	53½	4
Topgallant	10	23½	—	Topgallant	10	39½	2
Royal	9	15½	8, pole	Royal	7	30	1½
Bowsprit	30	30, outboard		Trysail mast	7	37	

Mast	Dia., inches	Length, feet	Head or pole, feet	Yard	Dia., inches	Length, feet	Arm, feet
*Jibboom, stem to cap	15	30		Spanker boom	10	54	1½, end
*Inner jibboom	15	18, outside		Spanker gaff	9	41½	5, end
*Outer jibboom	13	15		Main spencer gaff	6	21½	2, end
*Flying jibboom		2	7, pole				

NOTES

All dimensions are scaled from sail plan, ref. 11.
*Jibbooms are one stick, 72 feet whole length.
Masts buried below partners, fore 26½ ft.; main 24½ ft.; mizzen 27 ft.
Masts rake, ¾" fore; 1" main; 1¼" mizzen, per foot.
Mast spacing: knightheads to center of foremast, 54 ft.; thence to main, 78 ft.; thence to mizzen, 58 ft.; thence to taffrail, 51 ft.

Spar Plan #20

Commodore Perry, 1854 #1, 20 — packet, double topsails, Howes rig

Mast	Dia., inches	Length, feet	Head or pole, feet	Yard	Dia., inches	Length, feet	Arm, feet
Fore				Fore			
Topmast				Lower topsail			
				Upper topsail			
Topgallant				Topgallant			
Royal, sliding gunter				Royal, sliding gunter			
Main				Main			
Topmast				Lower topsail			
				Upper topsail			
Topgallant				Topgallant			
Royal, sliding gunter				Royal, sliding gunter			
Mizzen				Crojack			
Topmast				Lower topsail			
				Upper topsail			
Topgallant				Topgallant			
Royal, sliding gunter				Royal, sliding gunter			
Bowsprit				Spanker boom			
Jibboom				Spanker gaff			
Flying jibboom							

NOTES

"She has sliding gunter royal masts, so that in bad weather these can be sent down, and thereby ease her aloft, without interfering with the yards on the masts below" (Ref. 1).

Her lower masts are made. Topmasts and jibboom are hard pine.

No mast and spar dimensions are given.

Spar Plan #21

	Aurora, 1854 #28			packet, double topsails *Forbes* rig			
Mast	*Dia., inches*	*Length, feet*	*Head or pole, feet*	*Yard*	*Dia., inches*	*Length, feet*	*Arm, feet*
Fore				Fore			
Topmast				Lower topsail			
				Upper topsail			
Topgallant				Topgallant			
Royal				Royal			
Skysail				Skysail			
Main				Main			
Topmast				Lower topsail			
				Upper topsail			
Topgallant				Topgallant			
Royal				Royal			
Skysail				Skysail			
Mizzen				Crojack			
Topmast				Lower topsail			
				Upper topsail			
Topgallant				Topgallant			
Royal				Royal			
Skysail				Skysail			
Bowsprit				Spanker boom			
Jibboom				Spanker gaff			
Flying jibboom							

Notes

No mast dimensions are available. An article in *The Boston Daily Atlas*, 17 January 1855, states that the ship was fitted with Forbes rig with the topmasts fidded forward of the lower masts (Ref. 28).

Spar Plan #22

	Sierra Nevada, 1854 #1, 2			clipper, double topsails Howes rig			
Mast	*Dia., inches*	*Length, feet*	*Head or pole, feet*	*Yard*	*Dia., inches*	*Length, feet*	*Arm, feet*
Fore	37	87½	15	Fore	22	82	4½
Topmast	18½	50	9½	Lower topsail	19	72	4½
				Upper topsail	18	65	4½
Topgallant	12½	27	—	Topgallant	11½	49	3
Royal	11	18	10, pole	Royal	8	38½	2½
Main	40	91	15½	Main	23½	89	5
Topmast	19½	52	10	Lower topsail	20	77	5
				Upper topsail	19	70	5
Topgallant	13½	28	—	Topgallant	13	53	3
Royal	11½	18	12, pole	Royal	9½	42	2¼
Mizzen	30	80	12	Crojack	18	65	3¼
Topmast	15½	41	9½	Lower topsail	15	59½	3¼
				Upper topsail	14	53	3¼
Topgallant	10½	22	—	Topgallant	9½	39½	2¼
Royal	8½	14	8, pole	Royal	6½	30½	1¼

Mast	*Dia., inches*	*Length, feet*	*Head or pole, feet*	*Yard*	*Dia., inches*	*Length, feet*	*Arm, feet*
Bowsprit	34	20, outboard		Spanker boom		58	
Jibboom	19	20, outside		Spanker gaff		46	
Flying jibboom		16	5, end				

NOTES

All lower masts and bowsprit are made.
Topmasts and jibbooms are of hard pine.
Masts are "gracefully raked."

Spar Plan #23

Good Hope, 1855 #1 — freighter, double topsails, Howes rig, on fore and main only

Mast	*Dia., inches*	*Length, feet*	*Head or pole, feet*	*Yard*	*Dia., inches*	*Length, feet*	*Arm, feet*
Fore		77½		Fore		74	
Topmast		45		Lower topsail		65	
				Upper topsail		57	
Topgallant		24		Topgallant		44	
Royal		16	11, pole	Royal		33	
Main		81		Main		74	
Topmast		45		Lower topsail		65	
				Upper topsail		57	
Topgallant		24		Topgallant		44	
Royal		16	11, pole	Royal		33	
Mizzen		73		Crojack		58	
Topmast		37		Topsail		46	
Topgallant		20		Topgallant		34	
Royal		13	7, pole	Royal		26	
Bowsprit		18½, outboard		Spanker boom		48	
Jibboom		16, outside		Spanker gaff		38	
Flying jibboom		14	end	Main spencer gaff		22	

NOTES

Single topsail on mizzen mast.
Fore and main masts and bowsprit are made.

Spar Plan #24

Dione, 1857 #1 — freighter, double topsails, *See note below

Mast	*Dia., inches*	*Length, feet*	*Head or pole, feet*	*Yard*	*Dia., inches*	*Length, feet*	*Arm, feet*
Fore				Fore			
Topmast				*Lower topsail See note below			
				Upper topsail			
Topgallant				Topgallant			
Royal				Royal			
Main				Main			
Topmast				*Lower topsail See note below			
				Upper topsail			

Mast	*Dia., inches*	*Length, feet*	*Head or pole, feet*	*Yard*	*Dia., inches*	*Length, feet*	*Arm, feet*
Topgallant				Topgallant			
Royal				Royal			
Mizzen				Crojack			
Topmast				*Lower topsail See note below			
Upper topsail							
Topgallant				Topgallant			
Royal				Royal			
Bowsprit				Spanker boom			
Jibboom				Spanker gaff			
Flying jibboom							

NOTES

*The lower topsail yards are above the caps and are hung by chains to the topmasts, and the upper topsail yards hoist and lower in the usual style. The lower topsail yards are stationary.

Lower masts and bowsprit are made and hooped with iron.

Topmasts and jibboom are of hard pine.

Spar Plan #25

Emerald, 1855 #1 — freighter, double topsails *Howes rig

Mast	*Dia., inches*	*Length, feet*	*Head or pole, feet*	*Yard*	*Dia., inches*	*Length, feet*	*Arm, feet*
Fore	31	76	14	Fore	19	69	4
Topmast	16	41	8	Lower topsail	18	60	4
				Upper topsail	13	54	4
Topgallant	11	23	0	Topgallant	11	41	2½
Royal	9	14½	8, pole	Royal	8	32	1½
Main	31	76	14	Main	19	69	4
Topmast	18	43	8	Lower topsail	18	60	4
				Upper topsail	13	54	4
Topgallant	14	23	0	Topgallant	11	41	2½
Royal	9	15	8, pole	Royal	8	32	1½
Mizzen	25	72	12	Crojack	15	54	3
Topmast	13½	33	7	Topsail	11½	41	3
Topgallant	9	18	0	Topgallant	8	30	2
Royal	7	9	7, pole	Royal	6	22	1
Bowsprit	30	25, outboard		Spanker boom		45	
Jibboom		16, outside		Spanker gaff		32	
Flying jibboom		22	4, end				

NOTES

*Howes rig on fore and mainmast only.

Spar Plan #26

Sweepstakes, 1853 #2, 15 — clipper, Single topsails

Mast	*Dia., inches*	*Length, feet*	*Head or pole, feet*	*Yard*	*Dia., inches*	*Length, feet*	*Arm, feet*
Fore	36	86	15	Fore	21	78	5
Topmast	18	49	8-7 in.	Topsail	17	62½	5½
Topgallant*	12½	32½	5¼	Topgallant	11½	43½	2¾
Royal		20	—	Royal	9½	33	2
Skysail		12	7½, pole	Skysail	6	27	1½
Main	36	90	15	Main	1-10 in.	83	5
Topmast*	18½	51	9	Topsail	1-5 in.	67	5½
Topgallant	13	34	5½	Topgallant	12	48	2¾
Royal		21	—	Royal	9½	38	2
Skysail	10½	13	9, pole	Skysail	7½	31	1½
Mizzen	29	82	13	Crojack	17½	65	3-10 in.
Topmast	15	42	8	Topsail	12	50¼	4
Topgallant*	11	29	4½	Topgallant	9	35¼	2¼
Royal		16	—	Royal	6	26½	1
Skysail		10	6, pole	Skysail	5	21	1
Bowsprit	33 sq.	26, outboard		Spanker boom		56	
Inner jibboom		16, outside		Spanker gaff	10	47	
Outer jibboom		15¼		Swinging booms	10	49	
Flying jibboom	10	10, outside of wythe	5, end	Topmast stuns'l boom		50-11 in.	
				Tglt stuns'l boom		42	
				Royal stuns'l boom		34	

NOTES

*Topgallant mast is a separate spar with a masthead, all masts.
Royal and skysail masts and pole are a single spar, all masts.
A suit of sails spreads 1,000 yards of canvas, ref. 2.
Mast spacing, knightheads to foremast, 47 ft.; thence to mainmast, 67 ft.; thence to mizzenmast, 60 ft.; thence to taffrail, 50 ft.

Spar Plan #27

Stag Hound, 1850 #1, 2 — clipper, Single topsails

Mast	*Dia., inches*	*Length, feet*	*Head or pole, feet*	*Yard*	*Dia., inches*	*Length, feet*	*Arm, feet*
Fore	32½	82	13	Fore	20	72	4½
Topmast	16	46	9	Topsail	15	57	5
Topgallant	10	25	—	Topgallant	10	42	3
Royal	9	17	—	Royal	7	32	2
Skysail	8	13	7, pole	Skysail	6½	24¾	1½
Main	33	88	14	Main	22	86	4½
Topmast	17¾	51	9½	Topsail	17	68	5
Topgallant	12	28	—	Topgallant	15	53	3½
Royal	11	19	—	Royal	10½	42	2½
Skysail	10	15	9, pole	Skysail	7	32	1½
Mizzen	26	78	12	Crojack	16	60	4
Topmast	12½	40	8	Topsail	11½	48	4½
Topgallant	9	22	—	Topgallant	10	36	2½
Royal	8	16	—	Royal	7	27	1½

Mast	*Dia., inches*	*Length, feet*	*Head or pole, feet*	*Yard*	*Dia., inches*	*Length, feet*	*Arm, feet*
Skysail	7	11	6, pole	Skysail	6	22	1
Bowsprit	28½	24, outboard		Spanker boom	13	60	2½, end
Inner jibboom	16½	18, outside		Spanker gaff		44	5, end
Outer jibboom		15	5, end	Fore spencer gaff		25	2, end
Flying jibboom		18, outside of wythe	4, end	Main spencer gaff		25	2, end

NOTES

Masts rake 1¼" per foot, all masts.
Fore and main lower masts are made, fished on each side. Mizzenmast is a single spar.
Foremast is 29½ inches diameter at the truss band; mainmast is 30 inches diameter.
Bowsprit has 4½ inches steve to the foot.
Topmasts and jibbooms are of hard pine.
Tops are boarded solid, naval fashion.
Mast spacing, stem to center of foremast, 50 ft.; thence to mainmast, 67 ft.; thence to mizzenmast, 56 ft.; thence to sternpost, 42ft.

Spar Plan #28

Flying Fish, 1851 #1, 2 — clipper, Single topsails

Mast	*Dia., inches*	*Length, feet*	*Head or pole, feet*	*Yard*	*Dia., inches*	*Length, feet*	*Arm, feet*
Fore	38	82	14	Fore	21	70	4½
Topmast	18	46	8	Topsail	16	55	5
Topgallant	14	24	—	Topgallant	12	41	3
Royal	11	16	—	Royal	9	32	2
Skysail	9	12	6, pole	Skysail	7	23	1½
Main	40	88	15	Main	24½	80	4½
Topmast	19	49½	10	Topsail	17½	64	5
Topgallant	14	27	—	Topgallant	13	49	3½
Royal	12	18	—	Royal	10	39	2¼
Skysail	10	14	7, pole	Skysail	8	31	1½
Mizzen	26½	78	13	Crojack	16½	59	4
Topmast	15	38½	7½	Topsail	13	44	4½
Topgallant	12	21	—	Topgallant	10¾	34	2½
Royal	10	14	—	Royal	8	26	2
Skysail	8	10	5, pole	Skysail	6¼	20	1¼
Bowsprit	28	18, outboard		Spanker boom		58	
Jibboom	18	20, outside		Spanker gaff		40	5, end
Flying Jibboom		15	5, end	Fore spencer gaff		24	
Main spencer gaff		24					

NOTES

Fore and main lower masts are made and are fished on every square and hooped. The fishes extend two feet below the lower deck partners.
Bowsprit is 27 feet inboard, anchored by two sets of bitts.
When built *Flying Fish* was the most heavily masted Boston ship.
Masts rake 1¼" per foot.
Mast spacing—Stem to center of foremast, 52ft.; thence to mainmast, 62 ft.; thence to mizzenmast, 52 ft.; thence to sternpost, 44 ft.

Spar Plan #29

Challenge, 1851 #1, 2 — clipper, Single topsails

Mast	*Dia., inches*	*Length, feet*	*Head or pole, feet*	*Yard*	*Dia., inches*	*Length, feet*	*Arm, feet*
Fore	34	93	16½	Fore	22¾	81	4¾
Topmast	17¾	52	8¾	Topsail	15¾	66	6¼
Topgallant	12¼	34½	—	Topgallant	10¾	44	2½
Royal	8¼	22½	—	Royal	6¾	33	1¾
Skysail	6½	17	5, pole	Skysail	5½	22	1¼
Main	36	97	17	Main	24	90	5
Topmast	18½	54	9	Topsail	17	71	7
Topgallant	13	36	—	Topgallant	11½	47	3
Royal	8½	24	—	Royal	7¼	35	1¾
Skysail	7	19½	6, pole	Skysail	6½	23	1½
Mizzen	29	84½	13	Crojack	17½	70	3¼
Topmast	14½	42¾	7½	Topsail	13½	54½	4½
Topgallant	9¾	28½	—	Topgallant	8½	36½	2
Royal	6½	19½	—	Royal	6	27½	1½
Skysail	5½	15	4, pole	Skysail	5	18	1
Bowsprit	30	30, outboard		Spanker boom	13	60	3-2, end
Jibboom	17	20, outside		Spanker gaff	9	40	6½, end
Flying jibboom		15	5, end	Swinging booms	12	60	
Jib-a-jibboom		13, outside	3, end	Ringtail boom		30 (20, outboard)	
				Main spencer gaff	6	22	

Notes

Lower masts are made, fished on every square, and filled in under the hoops.

Tops are solid (naval fashion) and fit close to the eyes of the lower rigging. Fore top is 16½ ft. wide; main top 17–0 ft. wide; mizzen top 13½ ft. wide.

Fore and main lower yards are scarphed in the bunt.

Jib-a-jibboom is a separate spar.

With the lower studdingsails set on both sides the distance across the main yard is 160 feet. A suit of sails contains 12,780 yards of canvas.

Masts rake, 1⅛" foremast; 1¼" mainmast: 1½" mizzenmast.

Mast spacing, on deck; knightheads to center of foremast, 54½ ft.; thence to center of mainmast, 75½ ft.; thence to center of mizzenmast, 56½ ft.; thence to center of rudder stock, 43¼ ft.

Spar Plan #30

Sea Witch, 1846 #2, 14 — clipper, Single topsails

Mast	*Dia., inches*	*Length, feet*	*Head or pole, feet*	*Yard*	*Dia., inches*	*Length, feet*	*Arm, feet*
Fore		*55	12	Fore		67	3½
Topmast		39	7	Topsail		49½	2¾
Topgallant		20	—	Topgallant		37½	2
Royal		14	—	Royal		27	1½
Skysail		12½	3½, pole	Skysail		20½	1¼
Main		*59	12	Main		75	4
Topmast		41	7	Topsail		59	4
Topgallant		22	—	Topgallant		45	2
Royal		16	—	Royal		33	1½
Skysail		14	4, pole	Skysail		25	1¼

Mast	Dia., inches	Length, feet	Head or pole, feet	Yard	Dia., inches	Length, feet	Arm, feet
Mizzen		*50	9½	Crojack		60	3½
Topmast		31	4	Topsail		45	3
Topgallant		16½	—	Topgallant		33½	2
Royal		12	—	Royal		25½	1½
Skysail		9½	3, pole	Skysail		19	1¼
Bowsprit		23½, outboard		Spanker boom		50	1½
Inner jibboom		14, outside		Spanker gaff		37½	5
Outer jibboom		10		Main spencer gaff		22	
Flying jibboom		10	3, end				

NOTES

*Lengths of lower masts are measured from the deck.
*Masts rake, 1" fore; 1¼" main and mizzen, per foot.
Mast spacing, knightheads to center of foremast, 39 ft.; thence to mainmast, 57 ft.; thence to mizzen, 47ft.; thence to taffrail, 34½ ft.
No diameters are given.

Spar Plan #31

Flying Dutchman, 1852 #4 and see note below — clipper, Single topsails

Mast	Dia., inches	Length, feet	Head or pole, feet	Yard	Dia., inches	Length, feet	Arm, feet
Fore	30	81	14	Fore	19½	74	5
Topmast	16¾	47¼	8	Topsail	15	60	5½
Topgallant	11½	27	—	Topgallant	10½	42	3
Royal	7¾	18	—	Royal	6½	32	2½
Skysail	7	11¾	8, pole	Skysail	5½	24	1½
Main	31	85	14½	Main	21	81	5½
Topmast	16¾	48	8¼	Topsail	15½	66	6
Topgallant	11½	27½	—	Topgallant	10¾	46	3½
Royal	7¾	19	—	Royal	6¾	34	2½
Skysail	7	12	8, pole	Skysail	5½	25½	1½
Mizzen	23½	73	10½	Crojack	14	56	4
Topmast	12¾	38	6½	Topsail	11	47	4½
Topgallant	8¾	21¼	—	Topgallant	7¼	32	2½
Royal	5½	14	—	Royal	4	23	1½
Skysail	5	9	6, pole	Skysail	3½	16	1
Bowsprit	28	28½, outboard		Spanker boom	11	52	4½, end
Inner jibboom	16	14, outside		Spanker gaff	9	35	6, end
Flying jibboom	9	10, outside of wythe	4, end	Main spencer gaff	4	23	3 ½, end
Ringtail spar, outside		16					

NOTES

Dimensions are scaled from sail plan, ref. 4, and lifted from a printed spar list, origin undetermined. The two sources substantially agree.
Fore top is 14½ ft. wide; main top is 15-0 wide; mizzen top is 11½ ft. wide.

Spar Plan #32

Phantom, 1852 #1, 2 — clipper, Single topsails

Mast	*Dia., inches*	*Length, feet*	*Head or pole, feet*	*Yard*	*Dia., inches*	*Length, feet*	*Arm, feet*
Fore	36	82	14½	Fore	20	72	3½
Topmast	18	46	10	Topsail	16	56	4
Topgallant	13	25	—	Topgallant	12	40	2
Royal	10	16	—	Royal	9	32	2
Skysail	8	11	*	Skysail	7	25½	1½
Main	37	86	15	Main**	22	80	3½
Topmast	18	48	10	Topsail	17	61½	4
Topgallant	13	26	—	Topgallant	13	45	2
Royal	11	18	—	Royal	10	36	2
Skysail	9	12	*	Skysail	8	29	1½
Mizzen	27	76	13	Crojack	16	56	3
Topmast	15	40	8	Topsail	13	44	3½
Topgallant	10	22	—	Topgallant	10	33	2
Royal	8½	14	—	Royal	8	23	1¾
Skysail	7	9	*	Skysail	6	18	1¼
Bowsprit	35	20, outboard		Spanker boom			
Inner jibboom	18	18, outside		Spanker gaff			
Outer jibboom		18	5, end				
Flying jibboom		16, outside of wythe	3, end				

Notes

Fore and main lower masts are made, are of hard pine, doweled together, bolted, and hooped over all. Mizzenmast is a single spar. Bowsprit is 26 feet inboard of stem.

*No pole lengths are recorded.

**Fore and main yards are trussed seven feet from the masts.

Spar Plan #33

Coeur de Lion, 1854 #1, 2 — clipper, Single topsails

Mast	*Dia., inches*	*Length, feet*	*Head or pole, feet*	*Yard*	*Dia., inches*	*Length, feet*	*Arm, feet*
Fore		80		Fore		69	
Topmast		43½		Topsail		55	
Topgallant				Topgallant			
Royal				Royal			
Skysail				Skysail			
Main		82		Main		73	
Topmast		45		Topsail		58	
Topgallant				Topgallant			
Royal				Royal			
Skysail				Skysail			
Mizzen		76		Crojack		56	
Topmast		40		Topsail		46	
Topgallant				Topgallant			
Royal				Royal			
Skysail				Skysail			

Mast	*Dia., inches*	*Length, feet*	*Head or pole, feet*	*Yard*	*Dia., inches*	*Length, feet*	*Arm, feet*
Bowsprit				Spanker boom			
Inner jibboom				Spanker gaff			
Outer jibboom							
Flying jibboom							

NOTES

Fore and main lower masts are made in the "usual style."
Topgallant masts and above are pole masts.
Topmasts and jibbooms are of hard pine.
Other spars are in proportion.
Her rig (fitted with skysails) is based on a painting in ref. 2.
No additional details are recorded.

Spar Plan #34

Sword Fish, 1851 #2*, 4 — clipper, Single topsails

Mast	*Dia., inches*	*Length, feet*	*Head or pole, feet*	*Yard*	*Dia., inches*	*Length, feet*	*Arm, feet*
Fore	30	76	13½	Fore	17½	66	4¾
Topmast	16	44½	7⅔	Topsail	14½	54½	5¼
Topgallant	11	23	—	Topgallant	9½	37	2⅓
Royal	7	15	—	Royal	6½	28	1⅔
Skysail		10	6½, pole	Skysail	4¾	19½	1
Main	30	79	14	Main	19½	74	5
Topmast	16	46	8	Topsail	15½	61	5½
Topgallant	11	24	—	Topgallant	10	41½	2⅔
Royal	7	16	—	Royal	7	31½	1-10 in.
Skysail		10⅔	7, pole	Skysail	5	21½	1-2 in.
Mizzen	23½	69	11	Crojack	13½	52	4¼
Topmast	12	36	6¼	Topsail	10½	42	4½
Topgallant	8¼	18¾	—	Topgallant	6½	27¾	1¾
Royal	5½	12½	—	Royal	5	20½	1⅓
Skysail		8⅓	5½, pole	Skysail	4	13	0-9 in.
Bowsprit	38	25, outboard		Spanker boom	10	42	
Inner jibboom	16	18, outside		Spanker gaff	8	35	5½, end
Outer jibboom		12	8, end	Monkey gaff		15	
Flying jibboom				Swinging booms	9	42	

NOTES

These dimensions are taken from a contemporary masting list which is in agreement with the sail plan of *Sword Fish*.
The fore top is 14 feet wide; main top is 14½ feet wide; mizzen top is 11¼ feet wide.
*A painting in ref. 2 shows the vessel with Howes rig on the fore and main masts (Fig. 14), apparently a re-rig, date unknown.

Spar Plan #35

Golden Light, 1853 #1 — clipper, Single topsails

Mast	*Dia., inches*	*Length, feet*	*Head or pole, feet*	*Yard*	*Dia., inches*	*Length, feet*	*Arm, feet*
Fore	27	74	12½	Fore	16½	65	4
Topmast	15	42	7½	Topsail	14	52	4½
Topgallant		*	Topgallant			**	

Mast	*Dia., inches*	*Length, feet*	*Head or pole, feet*	*Yard*	*Dia., inches*	*Length, feet*	*Arm, feet*
Royal		*	Royal			**	
Skysail		*	Skysail			**	
Main	28	79	13	Main	19	74	4
Topmast	16	44	7½	Topsail	15½	60	4½
Topgallant		*		Topgallant		**	
Royal		*		Royal		**	
Skysail		*		Skysail		**	
Mizzen	24	69	10	Crojack	13	53	3½
Topmast	11½	32	6	Topsail	10	42	4
Topgallant		*		Topgallant		**	
Royal		*		Royal		**	
Skysail		*		Skysail		**	
Bowsprit	28	26, outboard		Spanker boom		45	
Jibboom		18, outside		Spanker gaff		38	
Flying jibboom		14		Main spencer gaff		24	

NOTES

Fore and main lower masts are made, bolted together, and hooped over all. Mizzenmast is a single spar.
Topmasts and jibbooms are of hard pine.
*Topgallant, royal and skysail masts are pole masts, in proportion.
**Topgallant, royal and skysail yards are in proportion.
Skysails were set flying, that is, they were sent aloft fully rigged and set only when suitable weather was encountered.

Spar Plan #36

Charmer, 1854 #1 — clipper, Single topsails

Mast	*Dia., inches*	*Length, feet*	*Head or pole, feet*	*Yard*	*Dia., inches*	*Length, feet*	*Arm, feet*
Fore		75		Fore		68	
Topmast		44		Topsail		56	
Topgallant		22		Topgallant		42	
Royal		14		Royal		32	
Skysail		9	6½, pole	Skysail		22	
Main		79		Main		72	
Topmast		47		Topsail		60	
Topgallant		24		Topgallant		46	
Royal		15		Royal		35	
Skysail		10	7½, pole	Skysail		25	
Mizzen		72		Crojack		56	
Topmast		37		Topsail		47	
Topgallant		20		Topgallant		36	
Royal		11½		Royal		28	
Skysail		8	6½, pole	Skysail		19	
Bowsprit	36	21, outboard		Spanker boom			
Inner jibboom		17, outside		Spanker gaff			
Outer jibboom		12					
Flying jibboom		10, outside of wythe	3, end				

NOTES

Fore and main lower masts are made, are of hard pine, and are hooped with iron.
Other spars are in proportion.

Bowsprit is gammoned with screw iron bands inside of the stem.
No additional details are recorded.
Skysail yards are rigged aloft.

Spar Plan #37

John Bertram, 1850 #1, 2 — clipper, Single topsails

Mast	*Dia., inches*	*Length, feet*	*Head or pole, feet*	*Yard*	*Dia., inches*	*Length, feet*	*Arm, feet*
Fore	29	76⅙	1½	Fore	18	68	4¼
Topmast	15	45½	8½	Topsail	14	53	4½
Topgallant	10 ½	24	—	Topgallant	10	39	3½
Royal	8	15	11, pole	Royal	7	29	2½
Skysail	See note			Skysail	6	23	1¼
Main	30	81⅙	13	Main	19	74	4¼
Topmast	16	47½	8½	Topsail	16	57	4½
Topgallant	11	26	—	Topgallant	11	42	3½
Royal	9	16	12, pole	Royal	8½	32	2½
Skysail	See note			Skysail	6½	26	1½
Mizzen	23	67¾	10	Crojack	14	54	3¼
Topmast	12	36	6½	Topsail	11½	41	3½
Topgallant	8½	20	—	Topgallant	9	31	2
Royal	6½	12	8, pole	Royal	7	21	1½
Skysail	See note			Skysail	5½	19	1
Bowsprit	28½	25½, outboard		Spanker boom	10½	32	4, end
Inner jibboom	16	20, outside		Spanker gaff		24	6, end
Outer jibboom		15	6, end				

Notes

The long poles, in lieu of measured skysail masts, indicate that skysails were set flying, that is, they were sent aloft fully rigged only when suitable weather was encountered.

Masts rake, 1¼" fore; 1½" main; 1¾" mizzen, per foot.

Spar Plan #38

Romance of the Sea, 1853 #1, 2 — clipper, Single topsails

Mast	*Dia., inches*	*Length, feet*	*Head or pole, feet*	*Yard*	*Dia., inches*	*Length, feet*	*Arm, feet*
Fore	31	78	14	Fore	19	72	5
Topmast	18	45	8½	Topsail	15	55	5½
Topgallant	13	23	—	Topgallant	9	38	3
Royal	8	15	8, pole	Royal	7	28	1½
Main	32	81½	14½	Main	21	83	5
Topmast	19	47	9	Topsail	16	63	5½
Topgallant	14	25	—	Topgallant	10	45½	3
Royal	9	16½	—	Royal	8	34	2
Skysail	5	13½	8, pole	Skysail	6	25	1½
Mizzen	28	76	12½	Crojack	15	57	4
Topmast	12	38	8	Topsail	11	45½	4½
Topgallant	10½	18	—	Topgallant	8	33	2½
Royal	7	12½	8, pole	Royal	6	24	1½
Bowsprit	30	22, outboard		Spanker boom		56	
Jibboom		22, outside		Spanker gaff		38	

Mast	Dia., inches	Length, feet	Head or pole, feet	Yard	Dia., inches	Length, feet	Arm, feet
Flying jibboom		17	5, end	Main spencer gaff		22	
				Fore spencer gaff		20	

NOTES

All lower masts and bowsprit are made, are of hard pine, doweled together, bolted and hooped over.
Topmasts and jibbooms are of hard pine.
Tops are extra wide and built solid of oak.

Spar Plan #39

Sovereign of the Seas, 1852 #1, 2 — clipper, Single topsails

Mast	Dia., inches	Length, feet	Head or pole, feet	Yard	Dia., inches	Length, feet	Arm, feet
Fore	41	89¾	16	Fore	22	80	5
Topmast	19	50	10	Topsail	17½	63	5½
Topgallant	14	27½	—	Topgallant	14	47	3½
Royal	11½	18	10, pole	Topgallant	8	37	2½
Main	42	92¾	17	Main	24	90	5
Topmast	19½	54	11	Topsail	19½	70	5½
Topgallant	14¾	30	—	Topgallant	15	53½	4
Royal	12	20	—	Royal	11	42	3
Skysail	10	14	8, pole	Skysail	9	35	2
Mizzen	34	82¾	14	Crojack	20	70	4
Topmast	16	43	9	Topsail	15	56	4½
Topgallant	11	24	—	Topgallant	11	43	3
Royal	9½	17	8, pole	Royal	7	32	2
Bowsprit	34	20, outboard		Spanker boom		61	2, end
Jibboom		15, outside		Spanker gaff		45	5, end
Flying jibboom		12	7, end	Main spencer gaff		24	2, end

NOTES

Mast lengths from heel to truck: fore, 185 ft.; main, 210 ft.; mizzen, 166 ft.

All lower masts are made, from the heads to the steps, each mast in five pieces, bolted and hooped together. The mast diameters at the truss bands are two inches less than at the deck.

The bowsprit is made; the outer pieces are of hard pine.

The bowsprit has four inches of steve to the foot.

Masts rake, ¾" fore, ⅞" main, 1⅛" mizzen, per foot.

Lower yards are single spars, not scarphed.

A suit of sails will spread between 11,000 and 12,000 yards of canvas.

Spar Plan #40

Elvira, 1855, #1 — freighter, Single topsails

Mast	Dia., inches	Length, feet	Head or pole, feet	Yard	Dia., inches	Length, feet	Arm, feet
Fore	31	51		Fore	19	67	
Topmast	16	42		Topsail	16	53	
Topgallant		23		Topgallant	10	38	
Royal		14	6, pole	Royal	5½	28	
Main	32	57		Main	19	70	
Topmast	16	44		Topsail	16	56½	

Mast	*Dia., inches*	*Length, feet*	*Head or pole, feet*	*Yard*	*Dia., inches*	*Length, feet*	*Arm, feet*
Topgallant		24		Topgallant	10	42	
Royal		15		Royal	5½	32¼	
Skysail		11	6½, pole	Skysail		26	
Mizzen	26	48		Crojack		54	
Topmast	13	33		Topsail		42	
Topgallant		17		Topgallant		32	
Royal		12	5½, pole	Royal		21	
Bowsprit		22, outboard		Spanker boom		41	
Jibboom		16, outside		Spanker gaff		31	
Flying jibboom		13	5, end	Fore spencer gaff		19	

NOTES

Masts rake, ⅞" fore; 1" main; 1¼" mizzen, per foot.
Steve of the bowsprit is 4¼" inches per foot.
Foremast, mainmast and bowsprit are made.
Topmasts and jibbooms are hard pine.

Spar Plan #41

Lightning, 1854 #1, 2 — clipper, Single topsails

Mast	*Dia., inches*	*Length, feet*	*Head or pole, feet*	*Yard*	*Dia., inches*	*Length, feet*	*Arm, feet*
Fore	37	86 65½ abv. deck	15	Fore	23	87	4
Topmast	18½	46	10½	Topsail	18½	72	6
Topgallant	13½	23	—	Topgallant	13	52	4
Royal	10½	15	7, pole	Royal	9½	40	1½
Main	38	90 72 abv. deck	16	Main	24	95	4
Topmast	19½	50	11	Topsail	19	72	6
Topgallant	14½	23	—	Topgallant	14	52	4
Royal	11½	15	—	Royal	10	40	2½
Skysail	8½	13	8, pole	Skysail	7	32	1
Mizzen	30	79 60 abv. deck	13	Crojack	19	72	3
Topmast	15½	40	8, pole	Topsail	15	52	4
Topgallant	11½	18	—	Topgallant	10	40	1½
Royal	8½	13	6, pole	Royal	7	32	1
Bowsprit	34	20, outboard		Spanker boom		50	
Inner jibboom	20	19, outside		Spanker gaff		38	6, end
Middle jibboom		14					
Flying Jibboom		15, outside of wythe		6, end			

NOTES

All lower masts and bowsprit are made, are of hard pine, doweled together and bolted, and hooped over all with iron.

All caps are of wrought iron: fore and main, one inch thick, 10 inches wide; others are in proportion.

"Heels of topmasts and topgallants, in way of their rigging, are curved out to correspond with the eyes of the rigging, so that these masts are wood-andwood in the doublings. Heels and heads of the masts are bound together below the caps by iron bands, set up with screws."

"Under the topmast trestletrees, around each masthead, let into the wood, is a massive iron screw-band, to give additional support to the crosstrees and all above them."

Topmasts and jibbooms are of hard pine.

The whole height of the mainmast, from deck to skysail truck, is 164 feet.

Masts rake, ¾" fore; ¾"main; 1¼" mizzen, per foot.

Mast spacing; 48 ft., stem rabbet to center line of foremast, at 20 ft. draft line; thence 75 ft. to c.l. of mainmast; thence 66½ ft. to c.l. of mizzenmast; thence 42 ft. to c.l. of rudder stock.

Spar Plan #42

War Hawk, 1855 #1 — clipper, Single topsails

Mast	*Dia., inches*	*Length, feet*	*Head or pole, feet*	*Yard*	*Dia., inches*	*Length, feet*	*Arm, feet*
Fore	32	75		Fore		68	
Topmast		44		Topsail		57	
Topgallant				Topgallant			
Royal				Royal			
Main	33	79		Main		72	
Topmast		47		Topsail		61	
Topgallant				Topgallant			
Royal				Royal			
Skysail, main only				Skysail, main only			
Mizzen	26	72		Crojack		56	
Topmast		41		Topsail		46	
Topgallant				Topgallant			
Royal				Royal			
Bowsprit		20, outside		Spanker boom			
Inner jibboom		17, outside		Spanker gaff			
Outer jibboom		14					
Flying jibboom		11					

Notes

All lower masts and bowsprit are made, are of hard pine, and are hooped with iron.
Masts and spars are in proportion.
No additional details are recorded.

Spar Plan #43

Bald Eagle, 1852 #1, 2 — clipper, Single topsails

Mast	*Dia., inches*	*Length, feet*	*Head or pole, feet*	*Yard*	*Dia., inches*	*Length, feet*	*Arm, feet*
Fore	35	85	15	Fore	21¾	80	4¼
Topmast	18	49	9	Topsail	18	63	5
Topgallant	14	26	—	Topgallant	12½	47	3
Royal	12	17	10, pole	Royal	9¼	26	2
Main	36½	90	15½	Main	23	87	4½
Topmast	18½	56	10, pole	Topsail	18¾	69	5
Topgallant	14½	27	—	Topgallant	12¼	53	3½
Royal	12½	18	—	Royal	11	43	2½
Skysail	10	14	12, pole	Skysail	9	31	1¼
Mizzen	28½	80	13	Crojack	17¾	62	4
Topmast	14	38½	8	Topsail	13¼	50	4½
Topgallant	10	21	—	Topgallant	10	38	3½
Royal	8½	14	8, pole	Royal	8	30	2
Bowsprit	28	22, outboard		Spanker boom	12	58	2, end
Jibboom	17½	22, outside		Spanker gaff	9	40	6, end
Flying jibboom							

Notes

All lower masts are made, the outside pieces extending the whole length from steps to the heads, doweled together, bolted and hooped over all.

The bowsprit is made, and has four inches steve to the foot.
Lower masts, topmasts, bowsprit, and jibbooms are of selected hard pine.
The studdingsail booms are one-half the diameter of the yards and are of the usual lengths.
A suit of sails will spread between 10,000 and 11,000 yards of canvas.

Spar Plan #44

	Winged Arrow, 1852 #1			clipper, Single topsails			
Mast	*Dia., inches*	*Length, feet*	*Head or pole, feet*	*Yard*	*Dia., inches*	*Length, feet*	*Arm, feet*
Fore		74		Fore		66	
Topmast		42		Topsail		53	
Topgallant				Topgallant			
Royal				Royal			
Main		80		Main		73	
Topmast		44		Topsail		59	
Topgallant				Topgallant			
Royal				Royal			
Skysail, main only				Skysail, main only*			
Mizzen		70		Crojack		53	
Topmast		32		Topsail		42	
Topgallant				Topgallant			
Royal				Royal			
Bowsprit		26, outboard		Spanker boom			
Jibboom		18, outside		Spanker gaff			
Flying jibboom							

NOTES

Fore and main lower masts are made.
Topgallant masts and above are pole masts.
Topmasts and jibbooms are of yellow pine.
Masts and spars are in proportion.
*Main skysail yard is rigged aloft.
No additional details are recorded.

Spar Plan #45

	Joseph Walker, 1850 #21			packet, Single topsails			
Mast	*Dia., inches*	*Length, feet*	*Head or pole, feet*	*Yard*	*Dia., inches*	*Length, feet*	*Arm, feet*
Fore		*57½	14	Fore		78	4½
Topmast		47	8	Topsail		65	5
Topgallant		26½	—	Topgallant		46	2½
Royal		18	4, pole	Royal		32	2
Main		*61	14	Main		80	5
Topmast		47	8	Topsail		65	5½
Topgallant		27½	—	Topgallant		46	2½
Royal		18	4, pole	Royal		32	2
Mizzen		*53	10½	Crojack		62	3½
Topmast		38	6½	Topsail		51	4½
Topgallant		23½	—	Topgallant		37	2
Royal		14	3, pole	Royal		25	1½
Bowsprit		33, outboard		Spanker boom		45	2½, end

Mast	*Dia., inches*	*Length, feet*	*Head or pole, feet*	*Yard*	*Dia., inches*	*Length, feet*	*Arm, feet*
Jibboom		19, outside		Spanker gaff		34	5½, end
Flying Jibboom		40½	2½, end	Trysail mast		42	
				Main spencer gaff		21	3, end

NOTES

*Lower mast lengths are above the deck. They are not whole lengths.

Masts rake, ¾" fore and main; 1" mizzen, per foot.

Mast spacing, on deck; 36 ft. stem to center of foremast; thence 64 ft. to mainmast; thence 47½ ft. to mizzenmast; thence 33 ft. to taffrail.

Dimensions are scaled from sail plan, ref. 21.

Spar Plan #46

Black Hawk, 1856 #4 — clipper, Single topsails

Mast	*Dia., inches*	*Length, feet*	*Head or pole, feet*	*Yard*	*Dia., inches*	*Length, feet*	*Arm, feet*
Fore		54*	12	Fore		69	3
Topmast		45	6½	Topsail		57	4½
Topgallant		24½	—	Topgallant		40	2
Royal		16	5½, pole	Royal		30	1¼
Main		57*	12½	Main		71	3
Topmast		45	7	Topsail		57	4½
Topgallant		24 ½	—	Topgallant		40	2
Royal		16	5½, pole	Royal		30	1¼
Mizzen		51*	10¼	Crojack		55	3
Topmast		30½	6	Topsail		44½	4
Topgallant		18	—	Topgallant		29	2
Royal		11½	5, pole	Royal		21	1¼
Bowsprit	27	25, outboard		Spanker boom		44½	2, end
Jibboom		12, outside		Spanker gaff		34	4½, end
Flying Jibboom		11½	3, end	Main spencer gaff		21½	2, end
				Monkey gaff		12	

NOTES

Dimensions are scaled from sail plan, ref. 4, except the bowsprit diameter which is given in the Table of Offsets. No diameters are shown in ref. 4.

*Lengths of lower masts are measurements above deck. They are not whole lengths.

Masts rake, ¾" fore; ⅞" main; 1⅛" mizzen, per foot.

Mast spacing, knightheads to center of foremast, 36 ft.; thence to mainmast, 62½ ft.; thence to mizzenmast, 47 ft.; thence to center of rudder post, 32½ ft.

Spar Plan #47

Invincible, 1851 #4, 6 — clipper, Single topsails

Mast	*Dia., inches*	*Length, feet*	*Head or pole, feet*	*Yard*	*Dia., inches*	*Length, feet*	*Arm, feet*
Fore	33	88¾	16	Fore	23	81¼	4¾
Topmast	18	51	8½	Topsail	17	65½	6¼
Topgallant	11	28¼	—	Topgallant	10	43¼	3
Royal	10	19	12½, pole	Royal	7	31¾	1¾
Main	36	93½	16½	Main	24	85½	5
Topmast	18½	53	9	Topsail	17½	69½	7

Mast	*Dia., inches*	*Length, feet*	*Head or pole, feet*	*Yard*	*Dia., inches*	*Length, feet*	*Arm, feet*
Topgallant	13	29¾	—	Topgallant	11	47¼	3¼
Royal	11	20	13½, pole	Royal	8	35	2¼
Mizzen	30	80½	12½	Crojack	19	65½	3½
Topmast	13½	42¼	7¼	Topsail	13	53	4 ½
Topgallant	11	23	—	Topgallant	8	36½	2
Royal	9	15¼	10½, pole	Royal	5½	27½	1¼
Bowsprit	35	30, outboard		Spanker boom	13	56	3½, end
Inner jibboom	17	18, outside		Spanker gaff	9	35	6, end
Outer jibboom	15	15		Main spencer gaff	7	25	3, end
Jib-a-jibboom	9	11, outside of wythe	3, end	Ringtail boom		30	
Swinging booms	12	54					

Notes

Bowsprit, jibbooms, spanker gaff, and other boom sizes are listed in ref. 6.
Mast and yard sizes are scaled from sail plan, ref. 4.
All lower masts are made with five pieces and filled in under the hoops, ref. 6.
Masts rake, 1⅛" fore; 1¼" main; 1½" mizzen, per foot.

Spar Plan #48

Union, 1850 #1 — freighter, Single topsails

Mast	*Dia., inches*	*Length, feet*	*Head or pole, feet*	*Yard*	*Dia., inches*	*Length, feet*	*Arm, feet*
Fore		71		Fore		64	
Topmast		40		Topsail		52	
Topgallant		—		Topgallant		37	
Royal		—	-, pole	Royal		30	
Main		75		Main		68	
Topmast		42		Topsail		55	
Topgallant		—		Topgallant		40	
Royal		—	-, pole	Royal		32	
Mizzen		67		Crojack		50	
Topmast		30		Topsail		39½	
Topgallant		—		Topgallant		28	
Royal		—	-, pole	Royal		22	
Bowsprit		27, outboard		Spanker boom		47	
Jibboom		24, outside		Spanker gaff		35	
Flying jibboom		18, outside of wythe, end included		Main spencer gaff		20	

Notes

Other masts are in proportion.
Union has pole topgallant and royal masts. No dimensions are given.

Spar Plan #49

	Champion of the Seas, 1854 #1, etc. See page 109			clipper, Single topsails			
Mast	*Dia., inches*	*Length, feet*	*Head or pole, feet*	*Yard*	*Dia., inches*	*Length, feet*	*Arm, feet*
Fore	40	65*	17	Fore	24⅓	88	5
Topmast	20	47	10	Topsail	19⅓	60	5½
Topgallant	15	26	—	Topgallant	13½	51	4
Royal	13	17	8, pole	Royal	9	37	2
Main	42	71*	17	Main	24½	95	5
Topmast	20	50	10	Topsail	19½	74	5½
Topgallant	15	27	—	Topgallant	13½	63	4
Royal	13	17	12, pole	Royal	9	42	3
Mizzen	36	61*	14	Crojack	20	74	4½
Topmast	16	42	9	Topsail	15	57½	5
Topgallant	12	24	—	Topgallant	9½	44	3½
Royal	10	15	8, pole	Royal	7	30	2
Bowsprit	40	22, outboard		Spanker boom		58-2 in.	
Jibboom	21	16, outside		Spanker gaff		42	
Flying jibboom		15	6, end	Main spencer gaff		22	

NOTES

All lower masts are made, are of hard pine, and are doweled together, bolted and hooped over all with iron.
Topmasts and jibbooms are of hard pine.
Caps are wrought iron. Doublings are nearly wood-and-wood (close together).
*Lengths of lower masts are measurements above deck. They are not whole lengths.
Mast spacing, foremast to mainmast, 74 ft.; mainmast to mizzenmast, 63 ft.; Masts rake ½" fore; ⅝" main; 1" mizzen, per foot.

Spar Plan #50

	Fearless, 1853 #2			clipper, Single topsails			
Mast	*Dia., inches*	*Length, feet*	*Head or pole, feet*	*Yard*	*Dia., inches*	*Length, feet*	*Arm, feet*
Fore				Fore			
Topmast				Topsail			
Topgallant				Topgallant			
Royal				Royal			
Main				Main			
Topmast				Topsail			
Topgallant				Topgallant			
Royal				Royal			
Mizzen				Crojack			
Topmast				Topsail			
Topgallant				Topgallant			
Royal				Royal			
Bowsprit				Spanker boom			
Jibboom				Spanker gaff			
Flying jibboom				Monkey gaff			

NOTES

Her description states that she carried single topsails and no sails above royals.
No details are recorded.
The painting in ref. 2 shows three jibs and a monkey gaff.

Spar Plan #51

	Alma, 1855 #1			freighter, Single topsails			
Mast	*Dia., inches*	*Length, feet*	*Head or pole, feet*	*Yard*	*Dia., inches*	*Length, feet*	*Arm, feet*
Fore		64		Fore		60	
Topmast		40		Topsail		47	
Topgallant		24		Topgallant		35	
Royal		16	5, pole	Royal		24	
Main		68		Main		60	
Topmast		40		Topsail		47	
Topgallant		24		Topgallant		35	
Royal		16	5, pole	Royal		24	
Mizzen		58		Crojack		47	
Topmast		32		Topsail		35	
Topgallant		17		Topgallant		24	
Royal		12	4, pole	Royal		13	
Bowsprit		14, outboard		Spanker boom		38	
Inner jibboom		16, outside		Spanker gaff		35	
Outer jibboom		13					
Flying jibboom		9					

NOTES

Other spars are in proportion.
Inner, outer and flying jibbooms are in one spar.

Spar Plan #52

	Winchester, 1851 #1			freighter, Single topsails			
Mast	*Dia., inches*	*Length, feet*	*Head or pole, feet*	*Yard*	*Dia., inches*	*Length, feet*	*Arm, feet*
Fore	32	84	13½	Fore	20	74	3½
Topmast	17	48	8½	Topsail	16	61	4½
Topgallant	12	26	0	Topgallant	11	44¾	3
Royal	10	18	12, pole	Royal	8	34 ½	2
Main	32	87	13½	Main	22	80	4
Topmast	17½	49½	8½	Topsail	17	66	4½
Topgallant	12½	26	0	Topgallant	12	48½	3
Royal	10½	18	13, pole	Royal	8½	38	2
Mizzen	26	80	12, pole	Crojack	16	60	3½
Topmast	13½	39	11½	Topsail	12	49	4
Topgallant	9½	20	0	Topgallant	8	35¾	2½
Royal	8	13	9, pole	Royal	6½	25½	1½
Bowsprit	33	30, outboard		Spanker boom			
Jibboom	19½	53*	3, end	Spanker gaff			
Flying jibboom		20, outside of wythe					

NOTES

Fore and mainmasts are fished on the sides and are closely hooped.
*This is total length of the jibboom. Divisions outside of the bowsprit cap are not noted
Flying jibboom is a separate spar.

Spar Plan #53

Racer, 1851 #2, 3 — clipper, Single topsails

Mast	*Dia., inches*	*Length, feet*	*Head or pole, feet*	*Yard*	*Dia., inches*	*Length, feet*	*Arm, feet*
Fore	31	84	19¾	Fore	20	74	4½
Topmast	17	47	8¾	Topsail	15	60	4¾
Topgallant	11	26	—	Topgallant	10	44	3⅓
Royal	9	16	*	Royal	8	34	2⅓
Main	33	88	15	Main	21	81	4¾
Topmast	17	48	9	Topsail	16	65½	5
Topgallant	11½	27	—	Topgallant	11	47	3½
Royal	9	17	*	Royal	8	36	2½
Mizzen	26	80	12½	Crojack	dimensions not given		
Topmast	14	39	7¾	Topsail	16	48	4¼
Topgallant	9	22	—	Topgallant	12	35	2¾
Royal	6½	13	*	Royal	8	24	1¾
Bowsprit	34	30, outboard		Spanker boom			
Jibboom	18	17, outside		Spanker gaff			
Flying jibboom		16	7, end				

Notes

All lower masts are single spars.
Topgallant and royal masts are pole masts.
*No pole lengths above the royals are given.
A suit of sails spreads 8,152 yards of canvas, ref. 2.
Other spars are in proportion.
Masts rake 1" fore; 1¼" main; 1½" mizzen, per foot.

Spar Plan #54

Winfield Scott, 1851 #1 — packet, Single topsails

Mast	*Dia., inches*	*Length, feet*	*Head or pole, feet*	*Yard*	*Dia., inches*	*Length, feet*	*Arm, feet*
Fore	30½	70	13½	Fore	19	68	
Topmast	16	45	8	Topsail		58	
Topgallant		—		Topgallant			
Royal		—	, pole	Royal			
Main	31	84	14	Main	20	74	
Topmast	16	46½	8½	Topsail		62	
Topgallant		—		Topgallant			
Royal		—	, pole	Royal			
Mizzen	25	67	11	Crojack	17	58	
Topmast				Topsail		48	
Topgallant		—		Topgallant			
Royal		—	, pole	Royal			
Bowsprit	30½	30, outboard	*	Spanker boom			
Jibboom				Spanker gaff			
Flying jibboom one spar							

Notes

Masts rake, 1" foremast; 1¼" mainmast; 1½" mizzenmast, per foot.

*Bowsprit is 22 ft. inboard.
Royal yards are rigged aloft.

Spar Plan #55

Polar Star, 1852 #1 — freighter, Single topsails

Mast	*Dia., inches*	*Length, feet*	*Head or pole, feet*	*Yard*	*Dia., inches*	*Length, feet*	*Arm, feet*
Fore		69		Fore		44	
Topmast		44		Topsail		54	
Topgallant				Topgallant			
Main		74		Main		64	
Topmast		44		Topsail		54	
Topgallant				Topgallant			
*Royal				Royal			
Mizzen		66		Crojack		48	
Topmast		30		Topsail		38	
Topgallant				Topgallant			
Bowsprit		26, outboard		Spanker boom			
Jibboom		26, outside		Spanker gaff			
Flying jibboom		16, outside of wythe					

NOTES

Masts rake, 1" fore; 1¼" main; 1½" mizzen, per foot.
No dimensions are recorded for spars left blank.
**Polar Star* carried only topgallants on fore and mizzen. Royal is on main only.
It is possible that mizzen topmast and topgallant are one pole.

Spar Plan #56

Antarctic, 1850 #1 — packet, Single topsails

Mast	*Dia., inches*	*Length, feet*	*Head or pole, feet*	*Yard*	*Dia., inches*	*Length, feet*	*Arm, feet*
Fore	30	80		Fore		72	
Topmast		46		Topsail		57	
Topgallant		—	-, pole	Topgallant		41	
Main	30	82		Main		72	
Topmast		47		Topsail		57	
Topgallant		—	-, pole	Topgallant		41	
Mizzen		74		Crojack		55	
Topmast		36		Topsail		44	
Topgallant		—	-, pole	Topgallant		32	
Bowsprit	30	30, outboard		Spanker boom		47	
Jibboom		16, outside		Spanker gaff		36	
Flying jibboom		14	6, end				

NOTES

No topgallant mast dimensions are given.

Spar Plan #57

	Young Turk, 1856 #1 bark			clipper, Single topsails Humphrey's rig			
Mast	*Dia., inches*	*Length, feet*	*Head or pole, feet*	*Yard*	*Dia., inches*	*Length, feet*	*Arm, feet*
Fore		51	9½	Fore		55	
Topmast		24	†	Lower topsail		47	
Stub topmast*		9½		Upper topsail		38	
Topgallant		14	—	Topgallant		32	
Royal		11	7, pole	Royal		26	
Main		54	9½	Main		55	
Topmast		23½	†	Lower topsail		47	
Stub topmast*		9½		Upper topsail		38	
Topgallant		14	—	Topgallant		32	
Royal		11	8, pole	Royal		26	
Mizzen		59	8½				
Topmast		25	—				
Topgallant		10	7, pole				
Bowsprit		16, outboard		Spanker boom		28	
Jibboom		16, outside		Spanker gaff		24	
Flying Jibboom		10	3, end				

NOTES

†No dimensions are given for the fore and main topmast heads.

*In this rig the lower topsail yard is set up to a stub topmast (fore and main masts only) which is installed immediately forward of the usual topmast and extends only from the lower top to the lower cap, 9½ feet. A description of this rig, taken from ref. 1, is included in Chapter 6. See also Fig. 6.9.

Masts and spars are in proportion.

No additional details are recorded.

Masts rake, 1" fore; 1¼" main; 1½" mizzen, per foot.

Spar Plan #58

	Racehorse, 1850 #1 bark			clipper, Double topsails Forbes rig			
Mast	*Dia., inches*	*Length, feet*	*Head or pole, feet*	*Yard*	*Dia., inches*	*Length, feet*	*Arm, feet*
Fore	21½	68	18½	Fore	13	53	3½
Topmast	12	40½	5	Lower topsail	10½	43½	3¼
				Upper topsail	8¼	34	2
Topgallant	7¼	14	—	Topgallant	6½	27	1½
Royal	6½	12	7, pole	Royal	4¾	21	1
Main	23	74-7 in.	19	Main	14¾	62	3½
Topmast	12	44	5½	Lower topsail	13	53	3½
				Upper topsail	10½	43½	3¼
Topgallant	8½	17	—	Topgallant	8¼	34	2
Royal	7¼	14	8½ pole	Royal	6½	27	1½
Mizzen	17	74–7 in.	10				
Topmast	9¾	30	—				
Topgallant	8	10	5, pole				
Bowsprit	23	25, outboard		Spanker boom		35	
Jibboom		20, outside		Spanker gaff		30	
Flying jibboom		12½	4, end	Main spencer gaff		15½	

NOTES

Masts rake alike, 1½" per foot.

Mast spacing, knightheads to center of foremast, 29½ ft.; thence to mainmast, 49 ft.; thence to mizzenmast, 28½ ft.; thence to taffrail, 21 ¼ ft.

Spar Plan #59

Aurelia, 1855 #1 bark — freighter, Single topsails Howes rig

Mast	*Dia., inches*	*Length, feet*	*Head or pole, feet*	*Yard*	*Dia., inches*	*Length, feet*	*Arm, feet*
Fore		63		Fore		54	
Topmast		36		Lower topsail		46	
				Upper topsail		44	
Topgallant		22		Topgallant		32	
Royal		20	, pole	Royal		24	
Main		65		Main		54	
Topmast		36		Lower topsail		46	
				Upper topsail		44	
Topgallant		22		Topgallant		32	
Royal		20	, pole	Royal		24	
Mizzen		65					
Topmast		49					
		—					
Topgallant							
Royal			, pole				
Bowsprit				Spanker boom			
Jibboom				Spanker gaff			
Flying jibboom							

NOTES

Spars without dimensions are in proportion.

Spar Plan #60

Sultana, 1850 #1 bark — clipper (marginal), Single topsails

Mast	*Dia., inches*	*Length, feet*	*Head or pole, feet*	*Yard*	*Dia., inches*	*Length, feet*	*Arm, feet*
Fore		58½		Fore		52	
Topmast		36		Topsail		40	
Topgallant		18½	—	Topgallant		29	
Royal		12	5, pole	Royal		20	
Main		60		Main		58	
Topmast		37		Topsail		45	
Topgallant		19½	—	Topgallant		33	
Royal		13	—	Royal		23	
Skysail		9	5½, pole	Skysail		16	
Mizzen		60					
Topmast		28	—				
Topgallant		14	5, pole				
Bowsprit		19, outboard		Spanker boom			
Jibboom		18, outside		Spanker gaff			

Mast	Dia., inches	Length, feet	Head or pole, feet	Yard	Dia., inches	Length, feet	Arm, feet
Flying jibboom		10, outside of wythe		Main spencer gaff			

NOTES

Jibboom and flying jibboom are separate spars. Jibboom is fitted to carry two jibs.

Fore and main topgallant masts and above, and mizzen topmast and above, are pole masts. No masthead sizes are given.

"We have not the dimensions of her spanker boom or gaff, nor her main spencer gaff, but these can easily be inferred from the size of the vessel."

Masts and spars are in proportion.

No additional details are recorded.

Masts rake, 1" fore; 1¼" main; 1½" mizzen, per foot.

Spar Plan #61

Trieste, 1857 #23, bark — freighter, Single topsails

Mast	Dia., inches	Length, feet	Head or pole, feet	Yard	Dia., inches	Length, feet	Arm, feet
Fore		*45½	11	Fore		59	3
Topmast		38½	7	Topsail		49	4
Topgallant		21	—	Topgallant		35	2
Royal		13	9, pole	Royal		26	1
Main		*50	11	Main		59	3
Topmast		38½	7	Topsail		49	4
Topgallant		21	—	Topgallant		35	2
Royal		13	—	Royal		26	1
Skysail		8½	4, pole	Skysail		19½	¾
Mizzen		*47	9½				
Topmast		27	—				
Topgallant		15	10, pole				
Bowsprit		23, outboard		Spanker boom		38	2, end
Jibboom		17½, outside		Spanker gaff		31	5, end
Flying Jibboom		9	1, end	Fore spencer gaff		18	2, end

NOTES

*Lower mast lengths are from the deck. They are not whole lengths.

Masts rake, 1" fore; 1" main; 1¼" mizzen, per foot.

Mast spacing on deck; 26 ft. stem to centerline of foremast; thence 51ft. to mainmast; thence 35½ ft. to mizzenmast; thence 24 ft. to taffrail.

Dimensions are scaled from ref. 23.

Mizzen topmast is a pole mast, divided with a stop for the rigging.

Spar Plan #62

Wildfire, 1853 #1, bark — clipper, Single topsails

Mast	Dia., inches	Length, feet	Head or pole, feet	Yard	Dia., inches	Length, feet	Arm, feet
Fore		51		Fore		52	
Topmast		31		Topsail		42	
Topgallant				Topgallant		33	
Royal				Royal		25	
Skysail				**Skysail		17	
Main		55		Main		58	
Topmast		33		Topsail		48	

Mast	Dia., inches	Length, feet	Head or pole, feet	Yard	Dia., inches	Length, feet	Arm, feet
Topgallant				Topgallant		39	
Royal				Royal		31	
Skysail				**Skysail		23	
Mizzen		55					
*Topmast		57					
Bowsprit				Spanker boom			
Jibboom				Spanker gaff			
Flying jibboom							

NOTES

*The mizzen topmast is a pole mast, "divided in the usual style," for the gaff topsail and stops for the rigging.
**Skysail yards are rigged aloft.
Masts rake, 1¼" fore; 1½" main; 1¾" mizzen, per foot.
No additional details are recorded.

Spar Plan #63

Henry Hill, 1855 #1 bark — clipper, Single topsails

Mast	Dia., inches	Length, feet	Head or pole, feet	Yard	Dia., inches	Length, feet	Arm, feet
Fore	23	45*	Fore			56	
Topmast	13	33½		Topsail		43	
Topgallant	8½	17		Topgallant		28	
Royal	6½	11	5, pole	Royal		20	
Main	23½	48*	Main			63	
Topmast	13½	36		Topsail		50	
Topgallant	9	19		Topgallant		34	
Royal	7	12		Royal		26	
Skysail		7	6, pole	Skysail		18	
Mizzen	18	45*					
Topmast	11	28					
Topgallant		14	6, pole				
Bowsprit		16, outboard		Spanker boom		33	
Jibboom		12, outside		Spanker gaff		30	
Flying jibboom		9	5, end				

NOTES

*Lengths of lower masts are measurements above deck. They are not whole lengths.
Fore and main lower masts are made.
Mizzen topmast, topgallant mast, and pole are one spar.
Topmasts and jibbooms are of hard pine.
Main skysail yard is rigged aloft.
No additional details are recorded.
Masts rake, 1⅛" fore; 1½" main; 1⅞" mizzen, per foot.

Spar Plan #64

Kremlin, 1850 #1 bark — freighter, Single topsails

Mast	Dia., inches	Length, feet	Head or pole, feet	Yard	Dia., inches	Length, feet	Arm, feet
Fore		65		Fore		53	

Mast	*Dia., inches*	*Length, feet*	*Head or pole, feet*	*Yard*	*Dia., inches*	*Length, feet*	*Arm, feet*
Topmast		36		Topsail		43	
Topgallant		20		Topgallant		31	
Royal		—	, pole	Royal		25	
Main		68		Main		58	
Topmast		38		Topsail		46	
Topgallant		21		Topgallant		34	
Royal		—	, pole	Royal		27	
Mizzen		66					
Topmast		25					
Topgallant		13					
Royal		—	, pole				
Bowsprit		23, outboard		Spanker boom		37	
Jibboom		15, outside		Spanker gaff		30	
Flying jibboom		10	7, end	Main spencer gaff		15	

NOTES

Sizes of royal masts and poles are not given.

Spar Plan #65

Alice Tainter, 1856 #22 bark	freighter, Single topsails

Mast	*Dia., inches*	*Length, feet*	*Head or pole, feet*	*Yard*	*Dia., inches*	*Length, feet*	*Arm, feet*
Fore		*47½	10	Fore		56	2½
Topmast		37½	6½	Topsail		46	3½
Topgallant		22½	—	Topgallant		31	1½
Royal		14½	6, pole	Royal		22	1
Main		*50½	10	Main		56	2½
Topmast		37½	6½	Topsail		46	3½
Topgallant		22½	—	Topgallant		31	1½
Royal		14½	6, pole	Royal		22	1
Mizzen		*49	8½				
Topmast		28	—				
Topgallant		12	6, pole				
Bowsprit		25, outboard		Spanker boom		47	2½, end
Jibboom		18½, outside		Spanker gaff		23½	3, end
Flying jibboom		12	6, end				

NOTES

*Lower mast lengths are from the deck. They are not whole lengths.

Masts rake, ¾" fore; ⅞" main; 1¼" mizzen, per foot.

Mast spacing on deck, 28½ ft. stem to centerline of foremast; thence 53 ft. to mainmast; thence 33½ ft. to mizzenmast; thence 27 ft. to taffrail.

Mizzen topmast is a pole mast, divided with stops for the rigging.

Dimensions are scaled from ref. 22.

Spar Plan #66

	Moneynick, 1854 #1 bark			packet, Single topsails			
Mast	*Dia., inches*	*Length, feet*	*Head or pole, feet*	*Yard*	*Dia., inches*	*Length, feet*	*Arm, feet*
Fore		54		Fore		54	
Topmast		32		Topsail		44	
Topgallant		17		Topgallant		33	
Royal		11		*Royal		25	
Main		58		Main		54	
Topmast		32		Topsail		44	
Topgallant		17		Topgallant		33	
Royal		11		*Royal		25	
Mizzen		57					
Topmast		25					
Topgallant		12					
Royal		7					
Bowsprit		15, outboard		Spanker boom			
Jibboom		12, outside		Spanker gaff			
Flying jibboom		10	4, end				

NOTES

*Royal yards are rigged aloft.
Other spars are in proportion.

Spar Plan #67

	Edisto, 1851 #1 bark			packet, Single topsails			
Mast	*Dia., inches*	*Length, feet*	*Head or pole, feet*	*Yard*	*Dia., inches*	*Length, feet*	*Arm, feet*
Fore		56		Fore		52	
Topmast		33		Topmast		44	
Topgallant		18	, pole	Topgallant		30	
Main		60		Main		52	
Topmast		33½		Topsail		44	
Topgallant		18		Topgallant		30	
Royal		11	, pole	Royal		22	
Mizzen		58					
Topmast		30					
Topgallant		11	9, pole				
Bowsprit		19, outboard		Spanker boom		35	
Jibboom		22, outside		Spanker gaff		30	
Flying jibboom		14	, end	Main spencer gaff		16	

NOTES

Fore topgallant and main royal pole sizes and flying jibboom end are not given.

Spar Plan #68

	Sea Bird, 1851 #1 bark			clipper, Single topsails			
Mast	*Dia., inches*	*Length, feet*	*Head or pole, feet*	*Yard*	*Dia., inches*	*Length, feet*	*Arm, feet*
Fore	19½	52½	9	Fore		50	
Topmast	11	31½	6	Topsail		40	
Topgallant		16	—	Topgallant		21	
Royal		10	5½, pole	Royal		*	
Main	19½	56½	9	Main		52	
Topmast	11	32½	6	Topsail		42	
Topgallant		17	—	Topgallant		23	
Royal		11	6½, pole	Royal		*	
Mizzen	15	56½					
Topmast		25	—				
Topgallant		11	7, pole				
Bowsprit	20½	18, outboard		Spanker boom		32½	
Jibboom		19, outside		Spanker gaff		23	
Flying jibboom		12	3, end	Main spencer gaff		12	

Notes

*No royal yards are noted in ref. l.
Masts rake, 1¼" fore; 1½" main; 1¾" mizzen, per foot.
Mizzen topmast, topgallant and pole are one stick, divided for the rigging.
No additional details are recorded.

Spar Plan #69

	Antelope, 1855 #1, 19 *barkentine			freighter, Double topsails Forbes rig (fitted with screw propeller)			
Mast	*Dia., inches*	*Length, feet*	*Head or pole, feet*	*Yard*	*Dia., inches*	*Length, feet*	*Arm, feet*
Fore	23	66	16	Fore		56	4
Topmast		50, one spar		Lower topsail		46	4
Upper topsail		36	3				
Topgallant				Topgallant		26	1½
Royal			, pole	Royal		20	1
Main	23	80					
Topmast		—					
Topgallant		—	, pole				
**Mizzen	17½	70					
Topmast		—	, pole				
Bowsprit	19	17, outboard		Boom foresail boom		38	
Jibboom		17, outside		Boom foresail gaff		28	
Flying jibboom		13	3, end	Main spencer gaff		28	
Spanker boom		36					
Spanker gaff		21					

Notes

*The term barkentine was unknown when this vessel was built. Terms such as demi-barque, three masted schooner, brig rigged and brigantern were variously used at the rig's inception.

**Mizzenmast is stepped on cabin floor.
Per ref. 1, "The rig is indescribable by any known rule."

Spar Plan #70

Belle, 1850 #1
brig

packet,
Single topsails

Mast	*Dia., inches*	*Length, feet*	*Head or pole, feet*	*Yard*	*Dia., inches*	*Length, feet*	*Arm, feet*
Fore		48		Fore	*no data on yards		
Topmast		32		Topsail			
Topgallant		—		Topgallant			
Royal		19		Royal			
Main		50		Main	*no data on yards		
Topmast		32		Topsail			
Topgallant		—		Topgallant			
Royal		19		Royal			
Bowsprit		15½, outboard		Trysail mast			
Jibboom		17, outside		Trysail boom		44	
Flying Jibboom		12		Trysail gaff		38	

Notes

**Belle* is described as "a full rigged brig" in ref. l.
Masts rake, 1" fore and main, per foot.

Spar Plan #71

Volante, 1852 #27
brig

freighter,
Single topsails

Mast	*Dia., inches*	*Length, feet*	*Head or pole, feet*	*Yard*	*Dia., inches*	*Length, feet*	*Arm, feet*
Fore		*41¾	8½	Fore		50	3
Topmast		32	5½	Topsail		43	3
Topgallant		17	—	Topgallant		29	2
Royal		11	7, pole	Royal		17½	1
Main		*45	9	Main		50	3
Topmast		32	5½	Topsail		43	3
Topgallant		18	—	Topgallant		29	2
Royal		11	8, pole	Royal		17½	1
Bowsprit		21, outboard		Spanker boom		50	2½
Jibboom		15, outside		Spanker gaff		34	5½
Flying jibboom (separate spar, includes 3 ft. doubling at end of jibboom)		11	2, end				

Notes

*Lower mast lengths are from the deck. They are not whole lengths.
Masts rake, ¾" fore; 1" main, per foot.
Mast spacing on deck; 29½ ft. from stem to foremast; thence 41 ft. to mainmast; thence 42 ft. to taffrail.
Topgallant masts are pole masts with stops for the rigging.
Dimensions are scaled from ref. 27.

Spar Plan #72

Belle, 1853 #1 — clipper,
brig — Single topsails

Mast	*Dia., inches*	*Length, feet*	*Head or pole, feet*	*Yard*	*Dia., inches*	*Length, feet*	*Arm, feet*
Fore		54		Fore		52	
Topmast				Topsail		42	
Topgallant				Topgallant		32	
Royal				Royal		22	
Main		58		Main		52	
Topmast				Topsail		42	
Topgallant				Topgallant		32	
Royal				Royal		22	
Bowsprit				Spanker boom			
Jibboom				Spanker gaff			
Flying jibboom							

Notes

Masts are raked in Baltimore clipper fashion. Royal yards are rigged aloft.
The sails are made so that their leeches form continuous lines from the head earings of the royals to the clews of the courses.
No additional details.

Spar Plan #73

Josephine, 1855 #24, 25 — freighter,
brigantine — Single topsails

Mast	*Dia., inches*	*Length, feet*	*Head or pole, feet*	*Yard*	*Dia., inches*	*Length, feet*	*Arm, feet*
Fore		*37½	7¾	Fore		45	2
Topmast		29	5	Topsail		36½	3
Topgallant		15½	—	Topgallant		25	1½
Royal		10	4, pole	Royal		18	¾
Main		*59½	7	Main			
Topmast		24	—	Topsail			
Topgallant		9	4, pole	Topgallant			
The hull was laid down as a side-wheel steamer, but machinery was never installed. Only fore and main masts were stepped. Ref. 24.							
Bowsprit		18½, outboard		Spanker boom		59½	2, end
Jibboom		10, outside		Spanker gaff		30	3, end
Flying jibboom		6	2, end				

Notes

*Lower mast lengths are from the deck. They are not whole lengths.
Masts rake, ¾" fore; 1" main, per foot.
Mast spacing on deck; 27 ft. from stern to center of foremast; thence 58½ ft. to mainmast; thence 45 ft. to rudder post.
Main topmast is a pole mast, divided with stops for the rigging.
Dimensions are scaled from ref. 25.

Spar Plan #74

Brenda, 1851 #1
Minna, 1851 #1
*topsail schooners

freighters,
square fore topsail

Mast	*Dia., inches*	*Length, feet*	*Head or pole, feet*	*Yard*	*Dia., inches*	*Length, feet*	*Arm, feet*
Fore	27	84		Fore	No data		
Topmast		**	pole	Topsail	No data		
Main	26	87					
Topmast		**	pole				
Bowsprit	27	23, outboard		Fore boom			
Jibboom				Fore gaff			
Flying jibboom				Main boom			
				Main gaff			

Notes

*Both vessels were built from the same plans.
**"She has long topmasts which imply that she may carry gaff topsails and topgallant sails."
Other spars are in proportion.

Spar Plan #75

Manhasset, 1852 #26
Plandome, 1852 # 26
schooners

freighters,
fore-and-aft rig

Mast	*Dia., inches*	*Length, feet*	*Head or pole, feet*	*Yard*	*Dia., inches*	*Length, feet*	*Arm, feet*
Fore		*71	9				
Topmast		**37	9, pole				
Main		*74	9				
Topmast		36	9, pole				
Bowsprit		23, outboard		Main boom		53½	2½, end
Jibboom		17½, outside		Main gaff		33	3, end
Flying jibboom		8½	3, end	Fore gaff		31½	2½, end

Notes

*Lower mast lengths are from the deck. They are not whole lengths.
**Fore topmast is divided at 27 feet with a stop for the rigging.
Masts rake 1¼" per foot.
Mast spacing on deck; 26 feet, stem to foremast; thence 35 feet to mainmast; thence 40 feet to taffrail.
Dimensions are scaled from ref. 26.
Both vessels were built from the same plans.

Glossary

Abaft. Toward the stern.
Adze. A sharp tool for use in smoothing timbers that have been axed or sawn.
Aft. The part of the vessel nearest the stern.
Aris Pieces. Long, thin strips of wood which are added to the width of fishes when the fishes do not completely fill the diameter of a mast. (See Figure 6.5, details G and H, item d.)
Axe. A sharp tool used for chopping or trimming rough wood.
Bale. A curved, hinged, iron strap used to retain a jibboom above a bowsprit.
Battens. Thin pieces of wood nailed to mastheads and the middle part of yards to prevent ropes from chafing.
Beard. To smoothly diminish the thickness of a board from one end to the other so that the thin end fairs into the surface of another board or timber.
Bed. The location of the greatest diameter of a bowsprit.
Bees. Pieces of wood secured to the sides of a bowsprit at its forward end through which headstays are led.
Bibbs. Brackets bolted to the forward side of the hounds, one under each trestletree, to support the weight of the mast assemblies above.
Board. Timber that is sawed to a thickness of one inch or less.
Bolsters. Short pieces of quarter-round softwood that are placed lengthwise across the crosstrees and fay against the sides of the masthead. Their sectional dimension is about one inch greater than the thickness of the trestletrees. Their purpose is to ease chafing of the eyes of the rigging (shrouds) which pass over them.
Bolts. Cylindrical pieces of iron, generally headed at one end. They are used to unite two or more pieces of wood into one assembly.
Boom. A spar fitted on the aft side of a mast. It is trimmed horizontally. It is used to extend the foot of a fore-and-aft sail.
Boom Irons. Rings that are installed at the quarters and ends of yards for the studdingsail booms to slide through.
Bow. The rounding part of a ship's side forward.
Bowsprit. A heavy spar projecting forward of the bow of a vessel from which the headsails are set.
Bowsprit Bitts. A pair of timbers set vertically in the deck which serve as partners and anchor the heel of the bowsprit.
Butt. The lower end of a tree next above the root.
Cant. Is to turn a timber over from one position to another.
Cap. A piece of wood with a square hole in the aft portion which is secured to the top of a masthead and has a round hole in the forward portion through which the upper masts can pass and be restrained in position.
Carlings. Fore and aft timbers which are installed between deck beams.
Center of Effort. The center of wind pressure on all the combined sails.
Center of Gravity. The point in a vessel where all its weights are in balance and may be assumed to act.
Chamfer. A beveled edge or corner.
Chapel. A groove along the seams of a built-up mast.
Check. A small split or crack in timber.
Chocks. Blocks of wood which fill empty spaces where they provide alignment, as in a top between the lower masthead and the heel of the topmast.
Coaks. Small surfaces of wood raised above the normal surface of a plank or timber. They are generally round or oblong and are used between two faying surfaces. The mating surface is tooled with corresponding mortises. Their purpose is to preclude two pieces of wood sliding out of position in reference to each other.
Complain. A mast complains when it gives a sign of weakness without apparent cause. or being sprung. Usually it is a sign of internal decay
Core. The central member of a made-mast. See **Spindle**.
Cranse Iron. A truss in the form of a yoke with a universal fitting. It retained the lower yard in its assigned location but allowed it to swivel as required to take advantage of the wind. Examples are shown in Figure 8.7.
Cross Trees. Light timbers placed athwartships across the trestletrees. They support the tops and spread the rigging.
Dasher Block. The small block at the peak of the spanker gaff. It serves the ensign halliard.
Depth. The distance from the head to the foot of a square sail.
Dolphin Striker see **Martingale Boom**.
Doubling. The overlap of the upper and lower ends of two masts, e.g. the topmast and the topgallant mast.
Drop. The distance from the head to the foot of a square sail. It is used especially in reference to the courses and topsails. See **Depth**.
Druxy. Whitish streaks or spots in a tree indicating decay.
Dub. To reduce or take away wood with an axe.
Eight Square. An octagonal cross section, all sides of which have equal dimensions. All angles are the same.
Ekeing. Lengthening to achieve a required dimension.
Elm. A wood with little liability of splitting due to its curly grain. Used extensively in making small parts such as caps, bibbs and bees.
Fair. A line or surface which is smooth and unblemished.
Fasces. A group of sticks bound tightly together to form a bundle.
Fay. To join two surfaces so closely and smoothly that no space appears between them.
Ferrule. A small iron hoop affixed to the extremities of yards, booms, etc.
Fid. A small iron rod which supports upper masts when fitted at the head of standing masts.

Fid Plate. A small, iron plate nailed on trestletrees to protect them against wear by a fid.

Fillings. Relatively thin strips of wood used to fill out the diameters of masts when the fishes do not contain sufficient material to do so.

Fir. In making masts it is the most useful wood in length, size and weight. It forms the core, or spindle, side trees, fillings, etc. of a made-mast. It is also used in making single tree masts.

Fish, Front. Also known as the paunch, it is a long piece of fir, hollowed on the inside to match the convexity of the mast and rounded on the back. It fastens to the fore side of a mast to provide additional strength.

Fishes, Side. Two long pieces of fir, placed on opposite sides of a made-mast to fill out its diameter.

Fore. The general front end of a vessel.

Fore and Aft. From stem to stern, or in that direction.

Forward. The foremost part of a vessel. Commonly pronounced "forrard."

Four Square. Having four sides of equal length and four angles, each of 90 degrees.

Funnel. A copper sleeve fitted over an upper mast for rigging to rest on when a mast is struck.

Furniture. The items and fittings such as bibbs, crosstrees, tops, trestletrees, caps, ironwork, etc., which collectively are necessary to outfit a finished mast. See Figure 6.7.

Gaff. A spar fitted on the aft side of a mast. Its outer end peaks up at an angle from horizontal. It is used to extend the head of a fore-and-aft sail.

Gammoning. A lashing, rope or iron, which holds the bowsprit down on the stemhead.

Given Diameter. The major diameter of a spar.

Graduating Batten. A length of wood marked in such a manner as to allow it to be used in determining the proportionate diameters at the various quarters of lower masts and yards. See Figure 4.3.

Half Sheave. A hole through the end of a spar, generally up and down, that acts as a fairlead for a line passing through it. See Figure 9.1.

Halliard. A rope or tackle used for hoisting yards.

Hance. A local curved rise made for strength or appearance.

Haunch. A sudden decrease in size.

Head of the Mast. That portion of a mast which is above the cheeks.

Heel. The lower end of a mast.

Heeling. The square part left at the lower end of a topmast or topgallant mast.

Hew. To trim off the rough edges of timber using an axe.

Hoops. Strong circles or bands of iron which are driven on made masts, yards and bowsprits.

Hoops, Clasp. Iron bands having a hinge in the middle and which fasten together with forelocks through mortises made in the ends.

Hoops, Mould. Iron bands that are not fashioned in a complete circle.

Hounds. The location next below the head of a mast where it is made more substantial in order to support the trestletrees and the top.

Housing. That portion of a mast below the deck between the partners and the mast step.

Jackstay. An iron rod along the top of a yard to which the sail is bent.

Jaws. Semi-circular pieces of wood built into the inner end of booms and gaffs to confine them to the masts on which they are installed. See Figure 9.2.

Keel. The long, longitudinal structure in the bottom of a vessel, on the centerline, which supports the transverse framing of the hull.

Keelson. Additional timber installed on top of the keel in order to provide increased strength.

Knuckle. The point of abrupt angular change of direction in a length of timber.

Leathering. Leather used to line the round holes in caps and the sweep of jaws in booms and gaffs.

Line. To scribe, or chalk, lines on timbers to denote locations, limits, etc.

Lining. Marking a surface with a straight line, e.g. a middle line, by stretching a cord between two points, coating it with chalk, and then snapping it vertically against the surface, leaving a straight imprint. (It is said that a curved line may be had by snapping the cord sideways. This, however, requires extensive practice.)

Made-Mast. A mast that is built of multiple sticks, fitted and bound tightly as a unit to achieve a desired diameter.

Martingale Boom. A short spar hanging from the end of the bowsprit. It fairleads the forestays aft to the bow.

Mast. An upright spar that is stepped in a vessel for the purpose of supporting the yards and sails.

Masthead. The square top portion of a standing mast above the hounds.

Monkey Gaff. A small spar carried high on the aftermost mast of a square rigged vessel. It flies the national ensign or other appropriate flag.

Mortise. A hollow made in one board to receive a tenon from another piece.

Moulded. The dimension of the depth of a piece of timber, contrary to sided which is the width.

Mould Hoop. An iron band that deviates from a true circular form.

Moulds. Patterns made to the shape and size of a mast.

Necking. The squared end of a round spar.

Nib. A small projecting end, generally tapered but never pointed.

Oak. A useful timber noted for its strength and durability.

Out of Winding. A piece of plank or timber that is not twisted out of a plane. A very desirable condition.

Parral. A fitting which holds a yard close to a mast.

Partners. The location of the greatest or given diameter in a mast. This location coincides with the deck structure which is made oversize to restrain a mast in its desired position.

Paunch. The front fish on a mast.

Piece. Generally a small piece of wood used to make up a deficiency (as at the hounds.)

Plain. The flat surface which remains between coaks.

Plane. A flat surface, as of a board or plank.

Plank. Sawn timber ranging from one inch thick to four inches.

Plumb. Perpendicular.

Plumb-Line. A cord, weighted at one end, which, when suspended unimpeded, will always hang perpendicularly.

Pokers. Tools with long handles and, at one end, pieces of flat plate, which are used to force heated hoops on a mast assembly.

Pole Mast. A long upper mast in one piece rather than being divided into separate individual sections.

Quarter Iron. A fitting restraining the inboard end of a studdingsail boom.

Quarters. A term given to some of the divisions on a mast, yard, etc., where the different diameters are set off for marking.

Rabbet. A small channel cut in the edge of one piece of wood to let into another.

Race. To scribe with a knife having a U-shaped blade.

Rake. The inclination of a vessel's masts from perpendicular.

Ram-Line. A long cord used to gain a straight middle line on a tree or mast. It is made fast at one end, then drawn taut and secured.

Rase. To strike a line on a surface.

Rim. A skirting around the upper side of tops.

Rind Gall. Local damage to a young tree which causes the bark or rind to grow within the substance of a tree.

Rounding. Finishing a mast from 16 square to round by dubbing off the protruding angles with a plane or drawknife.

Saddle. A semi-circular piece of wood secured to the upper side of a bowsprit to position the heel of the jibboom.

Scarph. Beveling the ends of two timbers so that, when overlapped and joined together, they appear as one piece.

Score. A groove in a piece of wood, such as the shell of a block, to admit or retain another piece of material.

Shake. A long, natural split in timber.

Shans. A situation found in yards where the direction of the grain of the wood exits the outer diameter near the yardarm.

Sheave. A solid wheel fitted in mortises cut in masts, yards, booms, blocks, etc., which rotates around a pin as an axis.

Sheet. A rope used to control a sail. In a square sail it is attached to the clews (lower corners).

Shift. The distance between two similar objects, e.g., two scarphs.

Sided. The dimension of the width of a piece of timber, contrary to the depth which is moulded.

Side Trees. The lower main pieces of a made-mast, outside of the spindle. They sometimes fill out the diameter of the mast. See Figure 6.3.

Sixteen Square. A square converted to 16 equal sides.

Slings. The middle portion of a yard.

Snape. To reduce the end of a piece of wood to a lesser substance but not to a point.

Spars. A general term applied collectively to the masts, yards, booms and gaffs of a vessel.

Spencer Gaff. A standing spar carrying a loose footed fore-and-aft sail on any mast of a square rigged vessel except the mizzen.

Spider Band. An iron band containing several eyebolts.

Spindle. The central sticks which make up the core of a made-mast.

Spring. A split or break in a spar.

Square. A sailor's term for the length of the yards of square sails.

Squared-Off. Said of trees which are trimmed of branches and roughly squared.

Squaring. Making one or more sides square by using a plumb-line.

Standing Mast. The lower mast, generally stepped on the keel and intended to be stationary for the life of the vessel.

Step. The pieces of timber installed on the keel into which the heel of a mast is stepped or positioned.

Stern. The after part of a vessel above the rudder post.

Steve (Stive). The angle that the bowsprit makes with horizontal.

Stick. A tree trunk that has been trimmed to a squared configuration.

Stops. Abrupt shoulders left in spars to prevent ropes from sliding past that location.

Strike. To draw a line or a circle.

Studdingsail Boom. A spar shipped as a portable extension to a yardarm. Pronounced "stunsl."

Sweep. The inner, semi-circular surface of jaws that engage the mast to which a boom or a gaff is secured.

Tackle. A purchase formed by two blocks connected by a rope.

Tail. The tapered lower portion of the cheeks which fairs into the substance of the mast.

Taunt. A term indicating lofty spars.

Tauntness. Refers to the height of masts.

Tenon. The end of a mast or spar, generally made square to fit a corresponding mortise.

Thawts. Pieces of plank or timber piled one upon another to support a mast laid horizontally above the ground so that it can be fashioned into its final configuration.

Thwartships. Crosswise of the decks; side to side.

Toggle. A small wooden pin.

Tongue. The tapered part of the lower end of a spindle or scarph.

Top. A platform projecting around the lower masthead. Its principal use is to extend the topmast shrouds athwartship.

Top-Rope. A rope passing through a block at the cap down to the deck. Its use is to hoist a topmast parallel with the standing mast into its assigned position.

Trace. The outline of an object.

Transverse. Athwartship or at a right angle to the keel.

Trees. Various sticks of which made masts are constructed, such as side trees, etc.

Trestletrees. The fore-and-aft structural members of a top.

Trimming. Cutting any piece of wood into the form and shape desired with an adze or axe.

Truck. A round or oval block of wood fitted on the extreme top of a mast assembly.

Truss. An iron yoke that retains the lower yards in their proper location. See **Cranse Iron.**

Trysail Mast. A short mast standing close abaft the mizzen mast.

Tye. The halliard by which a yard is hoisted.

Wedge. A tapered piece of hardwood used to confine an object, such as a mast, in a given position.

Whole Length. The total length of a standing mast, including the tenon at the heel and the tenon atop the masthead.

Winding. The condition of a board, plank or piece of timber when its edge is not in a straight plane.

Woolding. Several turns of rope, or an iron band, fitted tightly around a mast or bowsprit.

Wring. Twisting of a mast at the masthead. A frequent deficiency in standing masts.

Wythe (Withe). A boom-iron at the yardarm with a fitting through which the studdingsail boom runs in and out and which holds down the heel of the boom when rigged out. Pronounced "with."

Yard. A spar that crosses a mast horizontally, from which a square sail is set.

Yardarm. The outer quarter of a yard.

Yardarm Iron. The iron fitting which supports the outer portion of a studdingsail boom.

References

References are identified by number throughout the text and tables to provide direct access to the sources. ** Original or contemporary source. * Secondary source.

**1. *The Boston Daily Atlas, 1844–1857.* Descriptions of American-built clippers, packets and freighters (scantlings, size, arrangements, fittings, colors, etc., and masting dimensions). Reports by Duncan McLean, marine reporter. Microfilm available in the Library of Congress, Shelf NP4736.

**2. *American Clipper Ships, 1833–1858,* Howe and Matthews. Argosy Antiquarian, Ltd., 1967, New York. (Original was published in 1926.)

**3. Clipper ship descriptions, *The Boston Daily Evening Traveller,* 23 December 1850 to 21 September 1853.

**4. *Plans of Wooden Vessels—From a Fishing Smack to the Largest Clipper Ships, etc., from 1840 to 1869.* William H. Webb, New York, 1895.

**5. Clipper ship descriptions, *The New York Commercial Advertiser,* 22 May 1851 to 30 September 1854.

**6. Clipper ship descriptions, *The New York Herald,* 11 December 1850 to 28 December 1853.

**7. Description of the Largest Ship in the World, the new Clipper *Great Republic,* of Boston. Designed, built and owned by Donald McKay, 1853. Written by a sailor. Boston, Eastburn's Press. This is an on-site description of *Great Republic* noting pertinent details and containing five line drawings.

**8. "The clipper *Great Republic,* 1855, as rebuilt." *U.S. Nautical Magazine and Naval Journal,* April 1855, New York; Griffiths, Bates & Co., New York. (Article was reproduced in *Nautical Research Journal,* Vol. 21, No. 1, January, 1975.)

**9. "Clipper ship *Herald of the Morning*—Body Plan, Lines and Scantlings." *U.S. Nautical Magazine and Naval Journal,* Vol. III, No. 6, March 1856, Griffiths, Bates & Co., New York.

**10. *Clippers of the Port of Portsmouth and the Men Who Built Them, 1850–1855,* by Ray Brighton, Portsmouth Marine Society, Publication Five.

**11. "Ship *Ocean Monarch,* 1856, Sail Plan." Original plan is in the Livingston Library, Webb Institute, Glen Cove, New York.

**12. *Clipper ship* Red Jacket*—Lines and spar dimensions. U.S. Nautical Magazine and Naval Journal,* Vol. IV, No. 3, June 1856; Griffiths, Bates & Co., New York.

**13. "*Red Jacket,* 1853—Details of ship and description of launching." *Limerock Gazette,* November 5, 1853. (As reprinted in the *Rockland Gazette,* Rockland, Maine, circa 1920s.) Courtesy of the Rockland Museum, Rockland, Maine, and Stephen Hopkins.

*14. "*Sea Witch,* 1846—Sail Plan," in the book *Ship Models: How to Build Them* by Charles G. Davis, published by the Marine Research Society, Salem, Massachusetts, 1925.

*15. The Ship Model Builder's Assistant, Charles G. Davis, Salem, Massachusetts; Marine Research Society, 1926.

**16. *Young America,* port side off the bow; starboard side off the quarter. Photographs in the Peabody Museum, Salem, Massachusetts. Negatives Nos. 5232 and 11895.

**17. *Flying Cloud,* contemporary Lines Plan.

**18. "Barkentine *Antelope,* 1855." *U.S. Nautical Magazine and Naval Journal,* Vol. III, October 1855 to March 1856, published by Griffiths, Bates & Co., New York, 1856. Courtesy of the G. Blunt White Library, Mystic Seaport, Mystic, Connecticut.

19. *Brig* Antelope *in Boston Harbor,* 1855 painting by Fitz Hugh Lane, Mystic Seaport Museum, Mystic, Connecticut.

*20. *The Visual Encyclopedia of Nautical Terms Under Sail,* Basil W. Bathe, New York; Crown Publishers, Inc., 1978.

**21. "*Joseph Walker,* 1850. Lines and Sail Plans." Originals are in the Livingston Library, Webb Institute, Glen Cove, New York.

**22. "*Alice Tainter,* 1856. Sail Plan." Original plan is in the Livingston Library, Webb Institute, Glen Cove, New York.

**23. "Bark *Trieste,* 1857. Sail Plan." Original plan is in the Livingston Library, Webb Institute;, Glen Cove, New York.

*24. *William H. Webb, Shipbuilder,* by E. Dunbaugh and Wm. BuBois Thomas, Glen Cove, New York: Webb Institute of Naval Architecture, 1989.

**25. "Brigantine *Josephine,* 1855. Sail Plan." Original plan is in the Livingston Library, Webb Institute, Glen Cove, New York.

**26. "Schooners *Manhasset* and *Plandome,* 1852. Sail Plan." Original plan is in the Livingston Library, Webb Institute, Glen Cove, New York.

**27. "Brig *Volante,* 1852. Sail Plan." Original plan is in the Livingston Library, Webb Institute, Glen Cove, New York.

**28. "Forbes rig in *Aurora* and *Cornelius Grinnell.*" *The Boston Daily Atlas,* January 17, 1855.

*29. *Greyhounds of the Sea: The Story of the Clipper Ship,* by Carl C. Cutler, Halcyon House, New York, 1930.

**30. Models and Measurements (scantlings of 20 vessels designed by Donald McKay.) Notebook by Henry Hall, 1883, in possession of Penobscot Museum, Searsport, Maine. Note: This ledger-type notebook spells out the scantlings of 20 vessels designed and built by Donald McKay between 1846 and 1855. Of these, 12 vessels were clippers and 8 were packets. The information in this notebook is based on original papers of Donald McKay which were passed on to Mr. Hall by Mrs. McKay.

*31. *The Ship: An Illustrated History,* Bjorn Landstrom; Doubleday & Co., New York, 1961.

*32. *The Lore of Ships,* Tre Tryckare; Holt, Rinehart and Winston, New York, 1963.

**33. *Rudimentary Treatise on Masting, Mast-Making and Rigging of Ships,* Robert Kipping, N.A.; John Weale, 59 High Holborn, London, 1856.

**34. *Notes on Ships of the Past,* Robert Bennett Forbes; J.F. Cotter & Co., printers, Boston, 1888. Copy in possession of Clemson University, Clemson, North Carolina.

*35. *The Mariner's Dictionary,* Gersham Bradford, Weathervane Books, New York, 1952.

*36. *The Masting and Rigging of English Ships of War, 1625–1860,* James Lees; Naval Institute Press, Annapolis, Maryland, 1979.

**37. *The Merchant's and Mechanic's Assistant,* I.R. Butts, Boston, I.R. Butts & Co., 1856.

*38. *Textbook of Dendrology, Covering the Important Forest Trees of the United States and Canada,* William Morehouse Harlow, New York; McGraw-Hill Book Co., 1958. (Free Library of Philadelphia, Call No. 582H22lt.)

**39. *A Treatise on Masting Ships & Mast Making,* John Fincham, Landport, England.Third Edition, 1854. Reprinted by Conway Maritime Press, Ltd., 24 Bride Lane, Fleet Street, London, EC4Y8DR, 1982.

**40. Lloyd's Register of British and Foreign Shipping, 1855.

*41. *Know Your Woods,* Albert Constantine, Jr.; Charles Scribner's Sons, New York, edition of 1987.

*42. *The American-Built Clipper Ship, 1850–1856,* William L. Crothers; International Marine, Camden, Maine; a Division of McGraw-Hill Companies, 1997.

*43. *Live Oaking—Southern Timber for Tall Ships*, Virginia Steele Wood; Naval Institute Press, Annapolis, Maryland, 1981.

*44. *American Barns and Covered Bridges*, Eric Sloane; Dover Publications, Inc., Mineola, New York, 1954.

*45. *The Ship Modeler*, E. Armitage McCann, published by A.C. and H.W. Dickens, New York, 1929–1933.

**46. Miscellaneous correspondence between William H. Webb (shipyard) and various timber basins in the New York area, in possession of Webb Institute of Naval Architecture, Glen Cove, New York.

**47. *Steel's Elements of Mastmaking, Sailmaking and Rigging*, David Steel, 1794, reprint of 1932, by Edward W. Sweetman, New York.

**48. *A New Rig for Ships and Other Vessels*, Robert B. Forbes; printed by Weir and White, 11 Cornhill, Boston, 1849. Courtesy of Robert C. Leavitt, Miami, Florida.

**49. "Attaching Extra Topsail Yards to Vessels," Specification for U.S. Patent No. 16,650, 17 February 1857, Eber H. Linnell, Orleans, Massachusetts. Courtesy of Robert C. Leavitt, Miami, Florida.

**50. "Extra Yards to Topsails," Specification for U.S. Patent No. 11, 125, 20 June 1854, Frederic Howes, Yarmouth, Massachusetts. Courtesy of Robert C. Leavitt, Miami, Florida.

51. *The Oxford Companion to Ships and the Sea*, edited by Peter Kemp, published by the Oxford University Press, London, 1976.

*52. *Masting and Rigging the Clipper Ship and Ocean Carrier*, Harold A. Underhill, published by Brown, Son & Ferguson Ltd., Glasgow, 1958.

Index